BIRKHÄUSER

Oberwolfach Seminars
Volume 41

Christopher D. Hacon
Sándor J. Kovács

Classification of Higher Dimensional Algebraic Varieties

Birkhäuser

Authors:

Christopher D. Hacon
Department of Mathematics
University of Utah
Salt Lake City, UT 84112-0090
USA
e-mail: hacon@math.utah.edu

Sándor Kovács
Department of Mathematics
University of Washington
Box 354350
Seattle, WA 98195
USA
e-mail: skovacs@uw.edu

2000 Mathematics Subject Classification: Primary 14E30, 14D22; Secondary 14B05, 14C20, 14D20, 14E05, 14E15, 14F17, 14F18, 14J17

Library of Congress Control Number: 2010924096

Bibliographic information published by Die Deutsche Bibliothek
Die Deutsche Bibliothek lists this publication in the Deutsche Nationalbibliografie;
detailed bibliographic data is available in the Internet at <http://dnb.ddb.de>.

ISBN 978-3-0346-0289-1

Basel · Boston · Berlin
P.O. Box 133, CH-4010 Basel, Switzerland
Part of Springer Science+Business Media
Printed on acid-free paper produced from chlorine-free pulp. TCF ∞
Printed in Germany

ISBN 978-3-0346-0289-1
e-ISBN 978-3-0346-0290-7

9 8 7 6 5 4 3 2 1

www.birkhauser.ch

Preface

This book grew out of the Oberwolfach-Seminar *Higher Dimensional Algebraic Geometry* organized by the two authors in October 2008. The aim of the seminar was to introduce advanced PhD students and young researchers to recent advances and research topics in higher dimensional algebraic geometry. The main emphasis was on the minimal model program and on the theory of moduli spaces.

The authors would like to thank the Mathematishes Forshunginstitut Oberwolfach for its hospitality and for making the above mentioned seminar possible, the participants to the seminar for their useful comments, and Alex Küronya, Max Lieblich, and Karl Schwede for valuable suggestions and conversations.

The first named author was partially supported by the National Science Foundation under grant number DMS-0757897 and would like to thank Aleksandra, Stefan, Ana, Sasha, Kristina and Daniela Jovanovic-Hacon for their love and continuos support.

The second named author was partially supported by the National Science Foundation under grant numbers DMS-0554697 and DMS-0856185, and the Craig McKibben and Sarah Merner Endowed Professorship in Mathematics at the University of Washington. He would also like to thank Tímea Tihanyi for her enduring love and support throughout and beyond this project and his other co-authors for their patience and understanding.

Contents

Part I

Basics

CHAPTER 1

Introduction

1.A Classification

Algebraic geometry is the study of zeros of polynomial equations, for example the solution to r equations in n variables

$$P_1(x_1, \ldots, x_n), \ldots, P_r(x_1, \ldots, x_n).$$

The ultimate goal is to describe these solution sets as precisely as possible. Of course the set of solutions depends on where the variables x_i are allowed to take their values. For example, one could look for solutions of

$$P(x, y, z) = x^n + y^n - z^n$$

where x, y, z are integers. By Wiles' celebrated proof of Fermat's last theorem, there are very few solutions (the obvious ones) for $n \geq 3$. If instead we consider solutions which are complex numbers, then by the fundamental theorem of algebra, there are (infinitely) many solutions. In fact the solution set may be visualized as a cone over a Riemann surface of genus $(n^2 - 3n + 2)/2$.

In this book we will study solutions to polynomial equations over the complex numbers $\mathbb{C}$. In general it is very hard (or even impossible) to find explicit solutions to certain systems of polynomial equations (for example consider the simple case of one irreducible polynomial in one variable of degree ≥ 5). So we will focus instead on trying to achieve a qualitative understanding of the possible solution sets.

It is well known that a polynomial $P(x)$ of degree n in one variable has exactly n (complex) solutions when counted with the appropriate multiplicity. More generally, given a system of n polynomial equations in n variables

$$P_1(x_1, \dots, x_n), \dots, P_n(x_1, \dots, x_n),$$

if the set of solutions is finite, then there are at most $N_1 \cdots N_n$ solutions where $N_i = \deg P_i$. Notice however that even if we count the solutions with the appropriate multiplicity, their number might not add up to $N_1 \cdots N_n$. If however, we consider solutions at infinity (i.e., if we compactify $\mathbb{C}^n \subset \mathbb{P}^n$), then by Bezout's theorem, we have exactly $N_1 \cdots N_n$ solutions when counted with the appropriate multiplicity.

From now on we will then focus on infinite sets of solutions in projective space $\mathbb{P}^n$ (and hence on the solutions of the corresponding homogeneous polynomials). Every such set of solutions is the union of irreducible subsets, i.e., sets that are not the union of two proper subsets of solutions to polynomial equations. We will begin by considering smooth varieties (i.e., submanifolds of $\mathbb{P}^n$). This may seem to be a very restrictive point of view, however notice that for a general choice of the system of polynomial equations, the solution set is a smooth variety. Moreover, given any singular set of solutions, by Hironaka's celebrated result, one can "remove the singularities".

1.A.1 Curves

The first nontrivial case to consider is of course the case when the solution set is 1-dimensional, i.e., the case of **curves**. We begin with the following.

EXAMPLE 1.1 (Plane Curves). Let $X \subseteq \mathbb{P}^2$ be a projective plane curve defined by a homogeneous polynomial of degree d in three variables. It is easy to see that plane curves of degree d are parameterized by $\mathbb{P}^{\frac{d(d+3)}{2}}$. The continuous parameters are (up to multiplication by non-zero scalars) the coefficients of the defining equation of the curve. Note that the set of singular plane curves of degree d corresponds to a closed subset of $\mathbb{P}^{\frac{d(d+3)}{2}}$ and that two different equations might define isomorphic curves (for example any two smooth plane curves of degree 2).

REMARK 1.2. Note that this parameter space for plane curves is not a moduli space as isomorphic curves are parameterized by several different points. In other words, there is an action of the projective linear group by which one needs to take the quotient.

However, this is not the only possible way one can get isomorphic curves embedded differently in the same space. For instance, in $\mathbb{P}^n$ one has a rational normal curve for each degree up to n. These curves are all isomorphic to $\mathbb{P}^1$ but they are not projectively equivalent.

Finally note that most curves do not arise as plane curves.

We will address these issues in the sequel.

Regarding the problem that the same curve may be embedded in ways that are not projectively equivalent we need to ask:

Question 1.3. Is there a natural way to embed our varieties?

Recall that an embedding $f : X \hookrightarrow \mathbb{P}^n$ corresponds to a set of global sections $s_0, \dots, s_n$ generating a very ample line bundle $\mathscr{L}$ and separating points and tangent directions. Explicitly $\mathscr{L} = f^* \mathscr{O}_{\mathbb{P}^n}(1)$ and $s_i = x_i \circ f$, or if we forget about automorphisms of the ambient space for a moment, this brings up yet another question:

Question 1.4. How do we find (very) ample line bundles on a variety?

The problem is that our variety may not be given with a specific embedding, or even if it is given as a subvariety of a projective space, that given embedding may not be the natural one (if there is any such).

If a variety X, even if it is smooth, is given without additional information, it is really hard to find non-trivial ample line bundles, or for that matter, any non-trivial line bundles. There is practically only one possible series of non-trivial choices: tensor powers of the *canonical line bundle*, i.e., if X is smooth, we can take powers of ω_X, the determinant of the cotangent bundle $\Omega_X = T_X^*$. Of course, one may also consider the determinant of the tangent bundle, but note that this is simply the inverse of the canonical bundle and so it does not give an independent line bundle.

At this point it may seem to be an arbitrary choice between positive or negative powers, that is, whether we consider the canonical or the anticanonical bundle as our candidate for an ample line bundle. It turns out that positive powers are useful when studying "most" varieties (those of general type) and negative powers are useful when studying other varieties (eg. Fano varieties), and neither choice gives a satisfactory answer for certain other sets of varieties. However, in some measure the canonical bundle is useful when studying an even larger class of varieties. This will hopefully be apparent in what follows. Therefore, we will focus on the canonical bundle, but one may instead repeat the following questions for the anticanonical bundle (e.g., when studying Fano varieties).

Now that we have identified a natural line bundle, the next question is whether it carries useful information about the variety in question. Ideally, we would like it to be very ample or at least sufficiently ample that a tensor power will define an embedding into projective space. Therefore, we ask:

Question 1.5. Is the canonical bundle ample?

In the case of curves, the answer is quite straightforward: there are three types of behavior corresponding to the cases of genus $g = 0$, $g = 1$ and $g \geq 2$.

- $X = \mathbb{P}^1$: $\omega_X = \mathscr{O}_{\mathbb{P}^1}(-2)$ is anti-ample,
- X is an elliptic curve: $\omega_X \simeq \mathscr{O}_X$,
- X is any other smooth projective curve: ω_X is ample of degree $2g-2 = 2\chi(X, \omega_X)$.

In this last case $\omega_X^{\otimes 3}$ is always very ample and thus X can be embedded by the global sections of that line bundle:

$$\phi_3 : X \hookrightarrow \mathbb{P}^N = \mathbb{P}H^0(X, \omega_X^{\otimes 3}).$$

(Recall that if $s_0, \ldots, s_N$ is a basis of $H^0(X, \omega_X^{\otimes 3})$, then ϕ_3 is defined by $x \to [s_0(x) : \ldots : s_N(x)]$.) The obvious discrete invariant to consider now is the degree of this embedding, i.e.,

$$d = 3 \deg \omega_X = 6\chi(X, \omega_X).$$

Using Kodaira vanishing (3.37), Riemann-Roch and Serre duality we can compute N:

$$N + 1 = h^0(X, \omega_X^{\otimes 3}) = \chi(X, \omega_X^{\otimes 3}) = d + \chi(X, \mathscr{O}_X) = 5\chi(X, \omega_X).$$

REMARK 1.6. In fact, for most curves (i.e., $g \geq 2$), we can recover X from its canonical ring $R(\omega_X) = \oplus_{m\geq 0} H^0(X, \omega_X^{\otimes m})$. We have that $R(\omega_X)$ is finitely generated over $\mathbb{C}$ and $X \simeq \operatorname{Proj}(R(\omega_X))$. Notice moreover that $R(\omega_{\mathbb{P}^1}) \simeq \mathbb{C}$ and if X is an elliptic curve, then $R(\omega_X) \simeq \mathbb{C}[t]$ and so $R(\omega_X)$ is finitely generated for any curve X.

Therefore, we are interested in classifying smooth projective curves of degree 6χ in $\mathbb{P}^{5\chi-1}$, where $\chi = \chi(X, \omega_X)$. In this case the discrete invariant we need (in addition to $\dim = 1$) is the degree of the third pluricanonical embedding.

The fact that we found a fixed power of the canonical bundle that is very ample, and whose space of global sections has the same dimension independently of the actual curve, means that one may embed all the curves into a single projective space. Then a subset of an appropriate Hilbert scheme (cf. Chapter 12) parameterizes these curves. Unfortunately, this set is not yet a moduli space for several reasons; one of them is that the embedding of a curve into the ambient projective space is not unique. However, due to the nature of the construction the different embeddings are projectively equivalent, hence after factoring out by the action of the automorphism group of the projective space one does obtain a moduli space. Note that this is actually not as easy as it may seem as the above parameter set is not necessarily a subscheme, so taking the quotient is not a trivial operation. The same procedure can be followed using *stable curves* instead of smooth ones to obtain a compact moduli space which is a compactification of the moduli space of smooth curves. Finally, keep in mind that even though the construction works and we do end up with a moduli space and its modular compactification, there are still many unsolved problems regarding curves and their moduli spaces.

Given the success with curves, one would like to generalize these results to higher dimensions (the next case to consider is of course the case of surfaces, i.e., $\dim X = 2$).

EXERCISE 1.7. Show that if X is a curve of genus 0, then $X \simeq \mathbb{P}^1$.

EXERCISE 1.8. Show that all curves of genus 1 are isomorphic to a plane cubic.

EXERCISE 1.9. Show that not all curves of genus ≥ 1 are isomorphic. (In fact there is a 1-parameter family of curves of genus 1 and a $3g-3$ parameter family of curves of genus ≥ 2.)

EXERCISE 1.10. Show that for all curves X of genus ≥ 2, $|K_X|$ is basepoint-free and $|K_X|$ is very ample if and only if X is not hyperelliptic (i.e., if and only if X does not admit a finite morphism $f : X \to \mathbb{P}^1$ of degree 2).

EXERCISE 1.11. Show that X is a non-hyperelliptic curve of genus 3 if and only if X is isomorphic to a quartic curve. Deduce from this that not all curves of genus 3 are plane curves.

EXERCISE 1.12. Show that there is a 9-dimensional family of curves of genus 4.

EXERCISE 1.13. Show that any curve may be embedded as a smooth curve in $\mathbb{P}^3$ and has a birational map to a nodal curve in $\mathbb{P}^2$.

EXERCISE 1.14. Show that curves of genus $g \geq 2$ have finite automorphism groups.

1.A.2 Kodaira Dimension

The first question is how to generalize the invariant given by the genus g (or equivalently by $\chi(\omega_X) = g - 1$). If X is a smooth variety of dimension $n > 1$, then the natural choice might seem to be $p_g(X) = h^0(X, \omega_X)$. Notice however that there are surfaces such that ω_X is ample, but $p_g(X) = 0$. This problem can be solved by considering the plurigenera $P_m(X) = h^0(X, \omega_X^{\otimes m})$. If ω_X is ample, then $P_m(X)$ grows like a polynomial of degree n. In fact the following coarser invariant is useful.

Let X be a smooth projective variety and consider the rational map induced by a set of generators of $H^0(X, \omega_X^{\otimes m})$:

$$\phi_m : X \dashrightarrow \mathbb{P}^N = \mathbb{P}H^0(X, \omega_X^{\otimes m}).$$

DEFINITION 1.15. The **Kodaira dimension** of X is denoted by $\kappa(X)$ and defined as

$$\kappa(X) := \max\{\dim \phi_m(X) \text{ for } m > 0\}.$$

REMARK 1.16. Equivalently, one can define the Kodaira dimension $\kappa(X)$ as the transcendence degree of $R(\omega_X) = \oplus_{m\geq 0} H^0(X, \omega_X^{\otimes m})$ over $\mathbb{C}$ minus 1. Note that some authors define $\kappa(X) = -\infty$ if $P_m(X) = 0$ for all $m > 0$. We will often say that $\kappa(X) < 0$ in this case.

DEFINITION 1.17. X is of **general type**, if $\kappa(X) = \dim X$. In particular, if ω_X is ample, then X is of general type.

EXAMPLE 1.18. $\kappa(\mathbb{P}^n) = \dim \emptyset < 0$. In fact, for any Fano variety X, $\kappa(X) < 0$.

EXAMPLE 1.19. For curves we have (again) three cases:

- $\kappa < 0$: $X \simeq \mathbb{P}^1$ ($\chi < 0$),
- $\kappa = 0$: X is an elliptic curve ($\chi = 0$),
- $\kappa = 1$: X is a curve with ω_X ample ($\chi > 0$).

EXAMPLE 1.20. Let X be a uniruled variety. Then X does not admit any global pluricanonical forms and hence $\kappa(X) < 0$. It is conjectured that this characterizes uniruled varieties.

For more on the classification of uniruled varieties see [Mor87, §11] and [Kol96].

EXERCISE 1.21. Let $\mathscr{L}$ be a line bundle on a smooth projective variety. Show that if $\kappa(\mathscr{L}) = 0$, then there exists an integer $d > 0$ such that if $m > 0$, then $h^0(\mathscr{L}^{\otimes m}) = 1$ if and only if d divides m.

EXERCISE 1.22. Let $\mathscr{L}$ be a line bundle on a smooth projective variety. Show that there exists constants $a,\ b > 0$ such that

$$a \cdot m^{\kappa(\mathscr{L})} \leq h^0(\mathscr{L}^{\otimes m}) \leq b \cdot m^{\kappa(\mathscr{L})}$$

for all $m > 0$ sufficiently divisible. (Moreover, if $\mathscr{L}$ is big, the above inequality holds for any $m \gg 0$.)

1.A.3 Surfaces

In the case of surfaces (smooth complex projective varieties of dimension 2), there is a new phenomenon: given a smooth surface X and a point $x \in X$, we can produce a new surface $X' = \mathrm{Bl}_x X$ known as the **blow-up** of X at x which is obtained by replacing x by a **rational curve** $E \simeq \mathbb{P}^1 \subset X'$. More precisely, there is a morphism of smooth surfaces $\mu : X' \to X$ such that the induced morphism $X' - E \to X - x$ is an isomorphism and $\mu^{-1}(x) = E$ is a rational curve with $E^2 = K_X \cdot E = -1$ where K_X is a canonical divisor, i.e., a divisor corresponding to the line bundle ω_X. This process can be repeated infinitely many times with arbitrary choice of the point to be blown up, however it is clear that if Y is obtained from X by a finite sequence of blow-ups, then the geometry of X and Y share many common features. The most obvious one is that $f : Y \dashrightarrow X$ is a **birational map** (more informally X and Y are **birational**), i.e., that f induces an isomorphism between open subsets (namely the complement of the blown-up loci). Recall that equivalently we say that a rational map $f : Y \dashrightarrow X$ is birational if it induces an isomorphism between function fields $f^* : K(X) \to K(Y)$. It is also the case that the plurigenera are birational invariants, i.e., $P_m(X) = P_m(Y)$ for any $m \geq 0$ and any birational surfaces X and Y. It turns out that two surfaces X and Y are birational if and only if they are connected by a finite sequence of blow-ups, i.e., if there are sequences of blow-ups

$$X' = X_n \to X_{n-1} \to \cdots \to X_1 \to X \quad \text{and} \quad Y' = Y_m \to Y_{m-1} \to \cdots \to Y_1 \to Y$$

such that X' and Y' are isomorphic. It is therefore natural to study surfaces modulo birational equivalence.

There is another reason to study surfaces up to birational equivalence. Suppose that ω_X is ample, and that $f : Y \to X$ is the blow-up of X at a point $x \in X$ with exceptional curve $E = f^{-1}(x) \subset Y$. Then $K_Y \cdot E = -1$ and so ω_Y is not ample. One may still hope that for most surfaces X; there is a birational surface X' such that $\omega_{X'}$ is ample. A necessary condition is that $\kappa(X) = \dim X$, i.e., that X is of general type. As observed above, another necessary condition is that X' is not the blow-up of another surface. We have the following theorem of Castelnuovo.

Theorem 1.23. *Let $E \subset X$ be a -1 curve in a smooth projective surface so that $K_X \cdot E = E^2 = -1$ and $E \simeq \mathbb{P}^1$. Then there is a morphism of smooth surfaces $f : X \to Y$ such that f is the blow-up of Y at $y = f(E)$. We say that Y is obtained by* **blowing down** *E.*

So if X is a surface of general type containing a -1 curve E, then we simply replace X by the surface obtained by blowing down E. Since this decreases the second Betti number, after finitely many blow-downs, we may assume that X is **minimal**, i.e., that it contains no -1 curves. Unfortunately, it is not true that if X is minimal, then ω_X is ample, however we have the following.

Theorem 1.24. *If X is a surface with $\kappa(X) \geq 0$, then there exists a unique minimal surface $X_{\min}$ birational to X and a morphism $X \to X_{\min}$ given by blowing down finitely many -1 curves. Moreover $\omega_{X_{\min}}$ is semiample (that is $\omega_{X_{\min}}^{\otimes m}$ is basepoint-free for some $m > 0$).*

Note that $R(\omega_X) \simeq R(\omega_{X_{\min}})$ and as $\phi_m : X_{\min} \to \mathbb{P}H^0(X, \omega_{X_{\min}}^{\otimes m})$ is a morphism, $\omega_{X_{\min}}^{\otimes m}$ is the pull-back of an ample line bundle. It follows that $R(\omega_{X_{\min}})$ is finitely generated cf. (5.75). We have the following possibilities:

- $\kappa(X) = 2$. Then $\phi_5 : X_{\min} \to \mathbb{P}H^0(X, \omega_X^{\otimes 5})$ is a birational morphism that contracts curves E such that $E \cdot K_X = 0$, the image X_{can} has Du Val singularities and $X_{\text{can}} = \operatorname{Proj} R(\omega_X)$. See (3.1) for an example where X_{can} is not smooth.
- $\kappa(X) = 1$. Then for $m \gg 0$, $\phi_m : X_{\min} \to \mathbb{P}H^0(X, \omega_X^{\otimes m})$ is a morphism whose image is a curve $C = \operatorname{Proj} R(\omega_X)$ and whose general fiber is an elliptic curve (moreover $h^0(\omega_X^{\otimes m}) > 0$ for some $m \in \{4, 6\}$ and $h^0(\omega_X^{\otimes m}) \geq 2$ for some $m \leq 42$ cf. [Rei97, E.7.1]).
- $\kappa(X) = 0$. There are four possibilities for $X_{\min}$ known as Abelian surfaces, K3-surfaces, Enriques surfaces and bi-elliptic surfaces. In any event $\omega_{X_{\min}}^{\otimes m} \simeq \mathcal{O}_{X_{\min}}$ for $m \in \{4, 6\}$ cf. [Rei97, E.7.1].

If $\kappa(X) < 0$, then the minimal model is not unique (but it is well understood how two minimal models are related). For any minimal model $X_{\min}$, either $X_{\min} \simeq \mathbb{P}^2$ or we have a morphism $f : X_{\min} \to C$ to a curve C of genus $h^0(\Omega_X^1)$ whose general fiber is a rational curve. Notice conversely that if a surface admits a fibration by rational curves F, then as $K_X \cdot F = -2$, any section of $\omega_X^{\otimes m}$ must vanish along F and hence $R(\omega_X) \simeq \mathbb{C}$.

The above classification, gives a very precise description of surfaces with $\kappa(X) < 2$ and a qualitative understanding of surfaces of general type.

We would like to remark that a complete classification of surfaces of general type is not feasible. In fact, even if one fixes certain invariants such as the plurigenera $P_m(X)$ and the irregularity $q(X) = h^0(\Omega_X^1)$, it is very hard to determine if a surface with the given invariants exists; let alone to actually describe the moduli space of such surfaces. On the positive side we have the following.

Theorem 1.25. *Minimal surfaces of general type with fixed volume $K_{X_{\min}}^2$ belong to a bounded family.*

The idea behind the proof is as follows: since $\omega_{X_{\min}}^{\otimes 5}$ is generated and $\phi_5(X_{\min}) = X_{\text{can}}$, it follows that X_{can} is a surface in $\mathbb{P}^N$ of degree $25K_{X_{\min}}^2$. By a Hilbert scheme argument these belong to finitely many families. The corresponding minimal surface is then uniquely determined as the minimal resolution of the canonical model X_{can}.

In fact we can construct a moduli space for canonical (and for minimal) surfaces. Since we are able to use a fixed (bounded) power of the canonical bundle to embed all varieties under consideration, the problem of projectively non-equivalent embeddings (1.2) does not arise. Studying the appropriate Hilbert scheme then shows that the Hilbert points parameterizing the canonical models of surfaces of general type form a locally closed subscheme. This is of finite type because of the above mentioned boundedness and one may prove that it is also separated. The projective linear group of $\mathbb{P}^N$ acts on this Hilbert scheme naturally and the constructed locally closed subscheme is invariant under this action. The quotient, whose existence one still needs to prove, is then a moduli space for the canonical models. Since minimal models of surfaces of general type are unique in their birational class, we also obtain a moduli space for minimal surfaces of general type.

1.A.4 Higher dimensional varieties

In view of the classification results for surfaces, one might hope that for any variety X, we have that either

- $\kappa(X) \geq 0$. X has a (birational) minimal model $X_{\min}$ such that $R(\omega_X) \simeq R(\omega_{X_{\min}})$ is finitely generated, and $\omega_{X_{\min}}$ is semiample so that for some multiple $m > 0$, we have that $\phi_m : X_{\min} \to \mathbb{P}H^0(X, \omega_{X_{\min}}^{\otimes m})$ is a morphism whose image is $X_{\text{can}} \simeq \operatorname{Proj} R(\omega_X)$; or
- $\kappa(X) < 0$. X has a (birational) minimal model $X_{\min}$ such that there is a morphism $X_{\min} \to Z$ with $\dim Z < \dim X$ whose fibers are covered by rational curves.

Notice that if X is covered by rational curves, then $\kappa(X) < 0$. This gives a nice "geometric" (conjectural) characterization of varieties with $\kappa(X) < 0$.

Not surprisingly, in dimension ≥ 3 several new phenomena occur. First of all, there are 3-folds that do not admit a smooth minimal model as shown by the following example:

EXAMPLE 1.26. Let A be an abelian variety of dimension 3 and let X be its quotient by the $\mathbb{Z}_2$ action generated by the involution $x \to -x$. Then X has 2^6 isolated terminal singularities of index 2 so that $\omega_X^{\otimes 2} = \mathcal{O}_X$ is a line bundle. A resolution of X is given by blowing up these singularities. We obtain a morphism $f : X' \to X$ from a smooth variety X' such that $\omega_{X'}^{\otimes 2} = \mathcal{O}_{X'}(E)$ where E denotes the exceptional divisor. Suppose that $X_{\min}$ is a smooth minimal model of X', then as X is itself a minimal model, X and $X_{\min}$ are isomorphic in codimension 1 cf. (5.37). Consider a common resolution $p : Y \to X$ and $q : Y \to X_{\min}$. Then (as computed above) the multiplicity along E of $2K_Y - p^*(2K_X)$ is 1 whereas (since $X_{\min}$ is smooth and hence $\omega_{X_{\min}}$ is a line bundle) one sees that the multiplicity along E of $2K_Y - q^*(2K_{X_{\min}})$ is an even integer. This is impossible and so X has no smooth minimal model.

This forces us to allow varieties with mild singularities. It turns out that there are several fairly natural classes of singularities to work with, see Chapter 3. A second issue, is that even if X is birational to a smooth variety X' with ample canonical bundle, the rational map $X \dashrightarrow X'$ may not be a morphism.

Therefore, if we are to factor the rational map $X \dashrightarrow X'$ via a finite sequence of (hopefully) well-understood rational maps (that generalize blowing down -1-curves), we must allow maps that are not contraction morphisms and hence are not immediate generalizations of contractions of -1-curves.

Never-the-less, one may try and mimic the strategy that worked so well with surfaces.

- We start with a variety with mild singularities (in particular such that ω_X makes sense at least as a $\mathbb{Q}$-line bundle).
- We identify ω_X-negative curves and remove them "one at a time".

- After finitely many steps we obtain either a contraction morphism $X \to Z$ whose fibers are covered by rational curves or we obtain a minimal model $X_{\min}$ such that $\omega_{X_{\min}}$ is semiample.

It is in general quite hard to show that a line bundle $\mathscr{L}$ is semiample. A necessary and hopefully more tractable condition is that the line bundle be nef (i.e., $c_1(\mathscr{L}) \cdot C \geq 0$ for any curve $C \subset X$).

The first input for this ambitious strategy is provided by the following Contraction theorem (cf. [Mor82], [Kaw84], [Kol84], [KM98, §3]).

Theorem 1.27. *Let X be a variety with mild singularities such that ω_X is not nef. Then there is a curve C and contraction morphism $f : X \to Z$ such that $c_1(\omega_X) \cdot C < 0$ and if $D \subset X$ is a curve, then f contracts D if and only if D is numerically equivalent to a positive multiple of C.*

In other words, just as in the case of contracting -1-curves, the morphism f contracts some ω_X-negative curves.

We can now attempt to use the Contraction theorem to produce minimal models.

1.28 MINIMAL MODEL PROGRAM.

- **Start:** We begin with a variety X with mild singularities (eg. smooth).
- **Minimal Model:** If ω_X is nef, then we declare that X is a minimal model and we stop.

  Otherwise, by the Contraction theorem, there is a morphism $f : X \to Z$ such that $\varrho(X/Z) = 1$ and ω_X^* is relatively ample (equivalently $c_1(\omega_X) \cdot C < 0$ for any contracted curve $C \subset X$).
- **Mori Fiber Space:** If $\dim Z < \dim X$, then one can show that the fibers F are Fano varieties (i.e., ω_F^* is ample and $\varrho(F) = 1$) which are covered by rational curves. This can happen only if $\kappa(X) < 0$. This is considered a positive outcome and we stop.
- **Divisorial Contraction:** If $\dim Z = \dim X$ and f contracts a divisor, then Z has mild singularities. We may replace X by Z and repeat the process (go back to the Start). Since this step decreases the Picard number by 1, we may repeat this step only finitely many times.
- **Flip:** If $\dim Z = \dim X$ and f does not contract a divisor, then Z does not have mild singularities. In this case in fact $c_1(\omega_Z) \cdot C$ is undefined and so we can not even ask if ω_Z is nef. The only logical choice is to replace Z by the "smallest partial resolution" $f^+ : X^+ \to Z$ such that ω_{X^+} becomes a $\mathbb{Q}$-line bundle which is ample over Z. An easy computation shows that then

$$X^+ := \operatorname{Proj}_Z(R(\omega_Z)) \simeq \operatorname{Proj}_Z(\bigoplus_{m \geq 0} f_* \omega_X^{\otimes m}).$$

  The bad news is that $f^+ : X^+ \to Z$ exists if and only if $\oplus_{m \geq 0} f_* \omega_X^{\otimes m}$ is a finitely generated $\mathscr{O}_Z$ algebra, which is a hard problem closely related to our original questions (however note that we need only finite generation over $\mathcal{O}_Z$ rather than $\mathbb{C}$!). The good news is that if X^+ exists, then it is uniquely determined, it automatically has mild singularities and ω_{X^+} is ample over Z.

The rational map $X \dashrightarrow X^+$ is called a flip. It may be viewed as a surgery which is an isomorphism in codimension 1 and replaces some ω_X-negative curves by ω_{X^+}-positive curves.

At any rate, assuming that the flip $X \dashrightarrow X^+$ exists, we replace X by X^+ and repeat the procedure (go back to the Start). Note that $\varrho(X) = \varrho(X^+)$ and so it is not a priori clear that we can not repeat this step infinitely many times.

Assuming that flips always exist, one hopes that after iterating this procedure finitely many times, we either produce a Mori fiber space (i.e., a fibration $f : X \to Z$ whose fibers are covered by rational curves) or we remove all ω_X-negative curves and so we have that ω_X is nef, i.e., X is a minimal model (moreover we hope to show that ω_X is semiample). The goal is therefore to prove that:

- **Existence:** Flips always exist.
- **Termination:** There is no infinite sequence of flips.
- **Abundance:** Show that if X is a minimal model (i.e., it has mild singularities and ω_X is nef), then ω_X is semiample.

In dimension 3, Mori constructed flips in [Mor88] and termination is fairly straightforward. Abundance was established in [Miy88], [Kaw92a] and [KMMc94]. Recently, there has been much progress in dimension ≥ 4. By [Sho03] flips exist in dimension 4 and by [Sho06] minimal models for varieties with $\kappa(X) \geq 0$ exist in dimension 4. By [BCHM10], [HM08] and [YT06], it is known that:

- If X has mild singularities (eg. X is smooth), then $R(\omega_X)$ is finitely generated.
- If X is of general type and it has mild singularities, then it has a minimal model $X_{\min}$ such that $\omega_{X_{\min}}$ is semiample.
- Flips always exist.
- In many cases we can choose our sequence of flips so that it terminates.

Therefore, for varieties of general type, the picture is quite satisfactory as far as obtaining a minimal model is concerned. The next step is then to construct moduli spaces for these varieties. See (1.A.8) for more details.

Before discussing moduli theory, we will consider, at least briefly, the case of varieties with $\kappa(X) < \dim X$.

The first case to consider is the case when K_X is not pseudo-effective:

Theorem 1.29 [BDPP03]. *K_X is not pseudo-effective if and only if X is uniruled.*

In fact we will show (cf. 5.57) that if K_X is not pseudo-effective, then there is a birational map $X \dashrightarrow \bar{X}$ and a Mori fiber space $\pi : \bar{X} \to Z$. In particular a general fiber $\bar{X}_\eta$ of π is a Fano variety (so that $-K_{\bar{X}_\eta}$ is ample and in particular X_η is uniruled). Therefore, we can think of Fano varieties as being the building blocks of varieties with non pseudo-effective canonical class.

If K_X is pseudo-effective, then by the abundance conjecture mentioned above one expects that X has a minimal model $X \dashrightarrow \bar{X}$ and $\omega_{\bar{X}}$ is semiample. In particular, one expects the following:

Conjecture 1.30. *Let X be a smooth projective variety.*

(1.30.1) If K_X is pseudo-effective, then $\kappa(X) \geq 0$.

(1.30.2) If K_X is nef and $K_X \not\equiv 0$, then $\kappa(X) > 0$.

In higher dimensions, the main known case of the above conjecture is the following result.

Theorem 1.31 [Nak04]. *If $\kappa_\sigma(X) = 0$ (i.e., if K_X is pseudo-effective and if for any ample divisor H on X, $h^0(\mathcal{O}_X(mK_X + H))$ is a bounded function of m), then $\kappa(X) = 0$.*

REMARK 1.32. (1.30) in fact implies the abundance conjecture mentioned above.

1.A.5 Fano varieties

Fano varieties, are projective varieties with mild singularities such that ω_X^* is ample.

Since they have a natural ample line bundle; namely ω_X^*, there is hope to classify these varieties. For instance it is known that smooth Fano varieties of a given dimension form a bounded family.

Considering singular Fano varieties; it is conjectured (by Borisov-Alexeev-Borisov) that for any $\varepsilon > 0$, ε-log terminal Fano varieties of a given dimension form a bounded family. The condition on the singularities is necessary because of the following.

EXAMPLE 1.33. Let X_n be the cone over a rational normal curve of degree n, then X_n has a resolution $f : Y_n \to X_n$ such that Y_n is the $\mathbb{P}^1$-bundle over $\mathbb{P}^1$ given by the projectivization of the vector bundle $\mathcal{O}_{\mathbb{P}^1} \oplus \mathcal{O}_{\mathbb{P}^1}(-n)$ and f contracts a section E of self intersection $E^2 = -n$. An easy computation shows that $K_{Y_n} = f^*K_{X_n} - (1 - 2/n)E$ so that X_n is ε-log terminal for any $\varepsilon < 2/n$. Since $-K_{X_n}$ is ample, this shows that the ε-log terminal condition is necessary, even in dimension 2.

We remark that the above conjecture is known to be true in dimension 2 by [Ale94], but in dimension ≥ 3 it appears to be very difficult and only partial results are known (cf. [AB04], [BB92], [Bor01], [Kaw92b], [KMMT00]).

1.A.6 Varieties with Kodaira dimension 0

One of the main open problems is to understand and classify varieties of Kodaira dimension 0. The first natural problem is to show that these have a good minimal model, i.e., a minimal model such that $\omega_X^{\otimes m} \simeq \mathcal{O}_X$ for some $m > 0$. Hopefully one can then show that such m is determined by the dimension of X. In dimension 2 it suffices to choose $m = 12$ and in dimension 3, if X has canonical singularities, then $m = 2^5 \cdot 3^3 \cdot 5^2 \cdot 7 \cdot 11 \cdot 13 \cdot 17 \cdot 19$

suffices by [Kaw86] and [Mor86]. Note that even for $K3$ surfaces, the moduli space has infinitely many components.

A major difficulty in constructing and analyzing moduli spaces of varieties of Kodaira dimension 0 is that there is no natural choice of an ample line bundle, hence there is no natural embedding independent of the varieties. To remedy this issue the usual approach is to study *polarized varieties*, that is, pairs $(X, \mathscr{L})$ where $\mathscr{L}$ is an ample line bundle on X. Morphisms of polarized varieties are assumed to preserve the polarization (i.e., the choice of the ample line bundle). The downside of this approach is that in this way a given X appears many times, essentially for all the possible choices of $\mathscr{L}$. Of course, these choices may or may not give very different polarizations. This issue is closely related to the study of automorphisms of these varieties and can lead to connections with non-commutative geometry.

1.A.7 The Iitaka Fibration

Once the classification of Fano varieties, varieties of Kodaira dimension 0 and varieties of general type is settled, the next step is to study the fibrations that naturally occur as the end product of the minimal model program for varieties not of general type:

- fibrations whose general fiber is a Fano variety; and
- fibrations whose general fiber has Kodaira dimension 0.

Note that if K_X is not pseudo-effective, then it is known that there is a minimal model program that ends in a fibration of Fano type. If $\kappa(X) \geq 0$, then a weak version of the fibration of Kodaira dimension 0 is always known to exist:

1.34 IITAKA FIBRATION [Iit82, §11.6], [Mor87, 2.4], [Laz04a, 2.1.C]. Let X be a smooth projective variety with $\kappa(X) \geq 0$. Then there exists a birational model $X^\flat$ for X and a fibration $\varphi^\flat : X^\flat \to Y^\flat$ such that $Y^\flat$ is a smooth projective variety with $\dim Y^\flat = \kappa(X)$ and $\kappa(F^\flat) = 0$, where $F^\flat$ is the generic geometric fiber of $\varphi^\flat$. Furthermore, the birational class of $Y^\flat$ is uniquely determined by these properties.

1.A.8 Moduli spaces of varieties of general type

These are the varieties for which the techniques of the minimal model program yield the most information. One of the main applications of the minimal model program is the construction of moduli spaces for varieties of general type. Similarly to the case of curves, this is indeed the "general" case.

As explained in (1.28), the minimal model program is now known to work for varieties of general type. In other words, a variety of general type admits a *minimal model*. Then by the Basepoint-free Theorem 5.1, it also admits a *canonical model*. Yet in other words, we obtain that every variety of general type is birational to a canonically polarized variety, i.e., a variety with an ample canonical bundle.

Unfortunately this canonical model may be singular, but fortunately, the singularities forced by the canonical model are not worse than the ones we must allow in order to have a compact moduli space.

Here is a step-by-step description of the process of obtaining a canonical model of an arbitrary variety X:

- Apply Nagata's theorem [Nag62] to get $\hat{X}$, a proper closure of X.
- Apply Chow's lemma [Har77, Ex.II.4.10] if necessary to obtain $\bar{X}$, a projectivization of $\hat{X}$.
- Apply Hironaka's theorem [Hir64] to get $\widetilde{X}$, a resolution of singularities of $\bar{X}$.
- Apply the minimal model program (1.28) (including Mori fiber spaces) to get $X_{\min}$, a minimal model of $\widetilde{X}$. Restrict to the case $\kappa(X_{\min}) \geq 0$.
- The ring $R(\omega_{X_{\min}}) = \bigoplus H^0(X_{\min}, \omega_{X_{\min}}^{\otimes m})$ is finitely generated and we let $X_{\text{can}} = \operatorname{Proj} R(\omega_{X_{\min}})$. In this book we will focus on the case when $X_{\min}$ is of general type. In this case the map

$$X_{\min} \to X_{\text{can}} = \operatorname{Proj} \bigoplus_{m=0}^{\infty} H^0(X, \omega_{X_{\min}}^{\otimes m})$$

  is a birational morphism.

REMARK 1.35. Notice however, that if $X_{\min}$ is not of general type, then by the abundance conjecture, one expects that $\omega_{X_{\min}}$ is semiample so that for some $m > 0$, there is a morphism (the Iitaka fibration)

$$\phi_m : X_{\min} \to X_{\text{can}} \subset \mathbb{P}H^0(X, \omega_X^{\otimes m}).$$

By a result of Fujino and Mori cf. [FM00], the ring $\bigoplus H^0(X_{\min}, \omega_{X_{\min}}^{\otimes m})$ is isomorphic (after passing to appropriate truncations) to the ring $\bigoplus H^0(X_{\text{can}}, \mathcal{O}_{X_{\text{can}}}(m(K_{X_{\text{can}}} + \Delta)))$ of a klt pair[1] (X_{can}, Δ) where $K_{X_{\text{can}}} + \Delta$ is ample. Therefore, the above assumption (that $X_{\min}$ is of general type) is not too restrictive if one is willing to work with klt pairs.

Once we obtain the canonical model and settle on the class of singularities we allow, we have everything ready to construct a moduli space. In order to accomplish this we will need to prove a few properties. These will be explained in more detail in Chapter 13. Here we only mention them briefly.

As in the case of curves, we need to fix a discrete invariant. This will be the Hilbert polynomial of the canonical bundle of the canonical model. This is the analog of the *genus* of curves, in fact, it is equivalent to it in the case of curves.

We classify the canonical models of our original varieties by constructing a moduli space. This requires the following steps:

- Apply *boundedness* (13.A) to prove that it is possible to embed all canonical models with a fixed Hilbert polynomial into a fixed projective space.

[1] defined on page 37

- Apply *local closedness* (13.C) to prove that the locus of points in the Hilbert scheme (cf. Chapter 12) of this fixed projective space that parameterizes our canonical models forms a locally closed subset, so we can consider the action of the automorphism group of the ambient projective space on this set.
- Form the quotient of this set by the above group action [Vie91], [KeM97], [Kol97a].
- Apply *separatedness* (13.D) and *properness* (13.E) of the moduli functor to conclude that the resulting space is separated and proper.

CHAPTER 2

Preliminaries

2.A Notation

As usual, we denote by $\mathbb{Z}$, $\mathbb{Q}$, $\mathbb{R}$ and $\mathbb{C}$ the integers, the rational numbers, the real numbers, and the complex numbers. We let $\mathbb{Q}_{\geq 0} = \{x \in \mathbb{Q} | x \geq 0\}$, and we adopt similar conventions for $\mathbb{Z}$, $\mathbb{Q}$, $\mathbb{R}$ and $\mathbb{C}$ and ≥ 0, ≤ 0, > 0 and < 0. We will write $m \gg 0$ for any sufficiently big integer $m \in \mathbb{Z}$ and $0 < \varepsilon \ll 1$ for any sufficiently small positive real number $\varepsilon \in \mathbb{R}_{>0}$. The **round down** of $d_i \in \mathbb{R}$ is $\lfloor d_i \rfloor = \max\{m \in \mathbb{Z} | m \leq d_i\}$. The **round up** of $d_i \in \mathbb{R}$ is $\lceil d_i \rceil = -\lfloor -d_i \rfloor$ and the **fractional part** of $d_i \in \mathbb{R}$ is $\{d_i\} = d_i - \lfloor d_i \rfloor$.

Let k be an algebraically closed field of characteristic 0. Unless otherwise stated, all objects will be assumed to be defined over k. A **scheme** will refer to a scheme of finite type over k and unless stated otherwise a **point** refers to a closed point.

For a functor Φ, $\mathscr{R}\Phi$ denotes its derived functor on the (appropriate) derived category and $\mathscr{R}^i\Phi := h^i \circ \mathscr{R}\Phi$ where $h^i(C^\bullet)$ is the cohomology of the complex $C^\bullet$ at the ith term. Note that if $C^\bullet$ is a complex of sheaves, then $h^i(C^\bullet)$ is a sheaf as well. For a scheme X, $\mathbb{H}^i(X, \mathscr{C}^\bullet)$ denotes the **hypercohomology** of the complex of sheaves $\mathscr{C}^\bullet$, i.e., $\mathbb{H}^i = h^i \circ \mathscr{R}\Gamma$. Similarly, $\mathbb{H}^i_Z := h^i \circ \mathscr{R}\Gamma_Z$ where Γ_Z is the functor of cohomology with supports along a subscheme Z. Finally, $\mathcal{H}om$ stands for the sheaf-Hom functor and $\mathcal{E}xt^i := h^i \circ \mathscr{R}\mathcal{H}om$.

For a sheaf $\mathscr{F}$ on the scheme X, $H^i(X, \mathscr{F})$ denotes the usual i-th cohomology group of $\mathscr{F}$ which is, of course, the same as $\mathbb{H}^i(X, \mathscr{F})$ if one considers $\mathscr{F}$ as a complex. We will also adopt the commonly used convention that $h^i(X, \mathscr{F}) := \dim H^i(X, \mathscr{F})$. Even though this would seem to conflict with the notation used in the above definition of $h^i(C^\bullet)$, we believe that it will not cause any confusion.

2.B Divisors

Let X be a normal variety. In particular X is smooth in codimension 1.

A **prime divisor** P on X is a codimension 1 irreducible and reduced subvariety of X.

The group of **Weil divisors** on X, $\mathrm{WDiv}(X)$ is the torsion free $\mathbb{Z}$-module given by the set of all formal linear combinations $D = \sum d_i P_i$ of prime divisors on X with integral coefficients. We let $\mathrm{WDiv}_{\mathbb{Q}}(X) = \mathrm{WDiv}(X) \otimes_{\mathbb{Z}} \mathbb{Q}$ and $\mathrm{WDiv}_{\mathbb{R}}(X) = \mathrm{WDiv}(X) \otimes_{\mathbb{Z}} \mathbb{R}$.

A divisor $D \in \mathrm{WDiv}_{\mathbb{R}}(X)$ is **effective**, i.e., $D \geq 0$ if $D = \sum d_i P_i$ and $d_i \geq 0$ for all i. If $D = \sum d_i P_i \in \mathrm{WDiv}_{\mathbb{R}}(X)$, then we define $\|D\| = \sup\{|d_i|\}$.

A **principal divisor** is a divisor of the form $D = (f)$ where $f \in \mathbb{C}(X)$ is a rational function on X and (f) is the divisor given by the difference between the zeroes and poles of f (counted with the appropriate multiplicity).

The **support** of a divisor $D \in \mathrm{WDiv}_{\mathbb{R}}(X)$ is the subset $\mathrm{Supp}(D) \subset X$ given by the union of the components of $D = \sum d_i P_i$ appearing with coefficient $d_i \neq 0$.

For any $D \in \mathrm{WDiv}(X)$ the associated **Weil divisorial sheaf** is the $\mathcal{O}_X$-module $\mathcal{O}_X(D)$ defined by

$$\Gamma(U, \mathcal{O}_X(D)) = \{D + (f) \geq 0 \text{ where } f \in \mathbb{C}(U)\}.$$

A divisor $D \in \mathrm{WDiv}(X)$ is **Cartier** if it is locally principal, i.e. if there is a cover of X by open subsets U_j such that $D|_{U_j} = (f_j)$ for some rational function f_j on U_j. We let $\mathrm{Div}(X) \subset \mathrm{WDiv}(X)$ be the subgroup of Cartier divisors. If $\mathrm{Div}(X) = \mathrm{WDiv}(X)$, then we say that X is **factorial**. We let $\mathrm{Div}_{\mathbb{Q}}(X) = \mathrm{Div}(X) \otimes_{\mathbb{Z}} \mathbb{Q}$ and $\mathrm{Div}_{\mathbb{R}}(X) = \mathrm{Div}(X) \otimes_{\mathbb{Z}} \mathbb{R}$ and similarly for $\mathrm{WDiv}_{\mathbb{R}}(X)$. If $\mathrm{Div}_{\mathbb{Q}}(X) = \mathrm{WDiv}_{\mathbb{Q}}(X)$, then we say that X is **$\mathbb{Q}$-factorial**. If $D \in \mathrm{Div}(X)$, then $\mathcal{O}_X(D)$ is **invertible**. We will often refer to invertible sheaves as **line bundles**.

Two divisors $D_i \in \mathrm{WDiv}(X)$ are **linearly equivalent** $D_1 \sim D_2$ if $D_1 - D_2 = (f)$ is a principal divisor. Note that $D_1 \sim D_2$ if and only if $\mathcal{O}_X(D_1) \simeq \mathcal{O}_X(D_2)$. We let $|D| = \{D' \geq 0 | D' \sim D\}$ be the **linear series associated to** D. Note that $|D| \simeq \mathbb{P}H^0(\mathcal{O}_X(D))$.

If D is a Weil divisor and $V \subset |D|$ is a linear series, then the **base locus** of V is

$$\mathrm{Bs}(V) := \bigcap_{C \in V} \mathrm{Supp}(C)$$

(here we adopt the convention that $\mathrm{Bs}(V) = X$ if $V = \emptyset$). If $\mathrm{Bs}(V) = \emptyset$, we say that V is **basepoint-free**. If $V \neq \emptyset$, then the **fixed divisor** $\mathrm{Fix}(V)$ is the biggest divisor $F \geq 0$ such that $F \leq D'$ for all $D' \in V$. Note that $\mathrm{Supp}(\mathrm{Fix}(V))$ is the divisorial component of $\mathrm{Bs}(V)$. We will abuse notation and we will denote $\mathrm{Fix}(|D|)$ simply by $\mathrm{Fix}(D)$ (similarly for $\mathrm{Bs}(D)$ etc.). The divisor $\mathrm{Mob}(D) = D - \mathrm{Fix}(D)$ is the **mobile part** of D. We will say that D is **mobile** if $\mathrm{Fix}(D) = 0$.

Similarly, if $\mathscr{L}$ is a line bundle, we let $\mathrm{Bs}(\mathscr{L})$ be the intersection of the zero sets of global sections of $\mathscr{L}$ (if $H^0(X, \mathscr{L}) = 0$, we let $\mathrm{Bs}(\mathscr{L}) = X$). In particular, if $\mathscr{L} = \mathcal{O}_X(D)$, then $\mathrm{Bs}(\mathscr{L}) = \mathrm{Bs}(D)$.

Let $\mathscr{L}$ be a line bundle, then there is a rational map $\phi_{\mathscr{L}} : X \dashrightarrow \mathbb{P}H^0(X, \mathscr{L})$ which is defined on the complement of the base locus of $\mathscr{L}$. More concretely, let $s_0, \ldots, s_N$ be a basis of $H^0(X, \mathscr{L})$, then $\phi_{\mathscr{L}}(x) = [s_0(x) : \cdots : s_N(x)]$ for any $x \in X - \mathrm{Bs}(X)$. If $\mathscr{L}$ is generated by global sections, or equivalently when the corresponding divisor is basepoint-free, then $\phi_{\mathscr{L}}$ is a morphism and $\mathscr{L} \simeq \phi^*_{\mathscr{L}}\mathscr{O}_{\mathbb{P}^N}(1)$. In particular $\mathscr{L}$ is Cartier.

A line bundle $\mathscr{L}$ is called **semiample** if $\mathscr{L}^{\otimes m}$ is generated by global sections for $m \gg 0$. $\mathscr{L}$ is **very ample** if $\phi_{\mathscr{L}}$ is an embedding and it is **ample** if $\mathscr{L}^{\otimes m}$ is very ample for some integer $m > 0$. A line bundle $\mathscr{L}$ on X is called **big** if the rational map $\phi_{\mathscr{L}^{\otimes m}} : X \dashrightarrow \mathbb{P}^N$ is birational (on to $\phi_{\mathscr{L}^{\otimes m}}(X)$) for $m \gg 0$. Note that in this case $\mathscr{L}^{\otimes m}$ is not necessarily generated by global sections, so $\phi_{\mathscr{L}^{\otimes m}}$ is not necessarily defined everywhere.

Note that if $D \in \mathrm{Div}(X)$ and $C \subset X$ is a curve, then we can define the **intersection number** $D \cdot C = \deg(\mathscr{O}_X(D)|_C)$. If $D = \sum d_j D_j \in \mathrm{Div}_{\mathbb{R}}(X)$ and $C = \sum c_i C_i \in A_1(X, \mathbb{R})$, then we let $D \cdot C = \sum d_j c_i D_j \cdot C_i$. Two $\mathbb{R}$-Cartier divisors D_1 and D_2 are **numerically equivalent** $D_1 \equiv D_2$ if $(D_1 - D_2) \cdot C = 0$ for any irreducible curve $C \subset X$. We say that $D \in \mathrm{Div}_{\mathbb{R}}(X)$ is **nef** (i.e. **numerically eventually effective**) if $D \cdot C \geq 0$ for any irreducible curve $C \subset X$. Note that if D is semiample, then D is nef, but if E is an elliptic curve and P is a non-torsion element of $\mathrm{Pic}^0(E)$, then P is nef but not semiample.

The **Néron-Severi group** of X is the quotient of the group of Cartier divisors by numerical equivalences

$$N^1(X) = \mathrm{Div}(X)/\equiv .$$

$N^1(X)$ is a free abelian group of finite rank. The rank of this group is $\varrho(X)$, the **Picard number** of X. If $\mu : Y \to X$ is a resolution, then $\varrho(Y)$ is computed by the sum of $\varrho(X)$ and the number of exceptional divisors of μ cf. [AHK07].

If $D = \sum d_i D_i \in \mathrm{WDiv}_{\mathbb{R}}(X)$ and $D' = \sum d_i' D_i \in \mathrm{WDiv}_{\mathbb{R}}(X)$, then we let $D \wedge D' = \sum \min\{d_i, d_i'\} D_i$. The **round up** of a divisor is defined by $\lceil D \rceil = \sum \lceil d_i \rceil D_i$, the **round down** is defined by $\lfloor D \rfloor = \sum \lfloor d_i \rfloor D_i$ and the **fractional part** is defined by $\{D\} = \sum \{d_i\} D_i$. The **strata** of the support of a divisor D are given by the irreducible components of the intersections of the irreducible components of $\mathrm{Supp}(D)$.

If $\phi : X \dashrightarrow Y$ is a rational map and $D \in \mathrm{WDiv}_{\mathbb{R}}(X)$, then we let $\phi_* D \in \mathrm{WDiv}_{\mathbb{R}}(Y)$ be the **push-forward** of D. Recall that if D is a prime divisor, then $\phi_* D = 0$ if $\dim(\phi(D)) < \dim(D)$ and otherwise $\phi_* D = (\deg \phi|_D) D'$ where D' is the prime divisor corresponding to $\phi(D)$. If D is a prime divisor and $\phi_* D = 0$, we say that ϕ **contracts** D. If ϕ is birational and $D \in \mathrm{WDiv}_{\mathbb{R}}(Y)$, then we let $\phi^{-1}_* D$ be the **strict transform** of D. We say that a birational map $\phi : X \dashrightarrow Y$ **extracts no divisors** if ϕ^{-1} contracts no divisors.

EXERCISE 2.1. Let A, $B \in \mathrm{WDiv}_{\mathbb{Q}}(X)$. Show that if $A \sim_{\mathbb{R}} B$, then $A \sim_{\mathbb{Q}} B$. In particular if K_X is $\mathbb{R}$-Cartier, then it is $\mathbb{Q}$-Cartier.

EXERCISE 2.2. Show that if $D \geq 0$ is a Cartier divisor, then there is an integer $m > 0$ such that $\mathbf{B}(D) = \mathrm{Bs}(mD)$. See Section 2.E for the definition of $\mathbf{B}(D)$.

EXERCISE 2.3. Let A, $B \in \mathrm{Div}_{\mathbb{R}}(X)$.

(2.3.1) Show that $\lfloor A + B \rfloor \geq \lfloor A \rfloor + \lfloor B \rfloor$;

(2.3.2) show that $\lfloor f^*A \rfloor \geq f^* \lfloor A \rfloor$ and give an example where $\lfloor f^*A \rfloor \neq f^* \lfloor A \rfloor$ for some morphism $f : Y \to X$; and

(2.3.3) give an example where $A \sim_{\mathbb{R}} B$ but $\lfloor A \rfloor$ is not $\mathbb{R}$-linearly equivalent to $\lfloor B \rfloor$.

EXERCISE 2.4. Show: If $D \in \operatorname{Div}(X)$ is semiample, then $R(D) = \oplus_{m\geq 0} H^0(\mathscr{O}_X(mD))$ is finitely generated.

2.C Reflexive sheaves

Let X be a scheme and $\mathscr{F}$ an $\mathscr{O}_X$-module. The **dual** of $\mathscr{F}$ is $\mathscr{F}^* := \mathscr{H}om_X(\mathscr{F}, \mathscr{O}_X)$ and the m-th **reflexive power** of $\mathscr{F}$ is the double dual (or **reflexive hull**) of the m-th tensor power of $\mathscr{F}$:

$$\mathscr{F}^{[m]} := (\mathscr{F}^{\otimes m})^{**}.$$

Observe that since taking the dual is a (contravariant) functor, there is always a natural map,

$$\mathscr{F}^{\otimes m} \to \mathscr{F}^{[m]}.$$

This map is injective if and only if $\mathscr{F}^{\otimes m}$ is torsion free. $\mathscr{F}$ is called a **reflexive $\mathscr{O}_X$-module** or simply **reflexive** if $\mathscr{F} = \mathscr{F}^{[1]} = \mathscr{F}^{**}$.

As we declared earlier, a **line bundle** on X is an invertible $\mathscr{O}_X$-module. A **$\mathbb{Q}$-line bundle** $\mathscr{L}$ on X is a reflexive $\mathscr{O}_X$-module of rank 1 such that one of its reflexive powers is a line bundle, i.e., there exists an $m \in \mathbb{Z}_{>0}$ such that $\mathscr{L}^{[m]}$ is a line bundle. The smallest such m is called the **index** of $\mathscr{L}$.

On a normal variety (or more generally on an S_2 variety), Weil divisorial sheaves are in one to one correspondence with reflexive $\mathscr{O}_X$-modules of rank 1. In particular, a reflexive $\mathscr{O}_X$-module of rank 1 is a line bundle on $X \setminus \operatorname{Sing} X$.

A Weil divisor D on X is a **Cartier** divisor, if its associated Weil divisorial sheaf, $\mathscr{O}_X(D)$ is a line bundle. This is equivalent to the definition we gave earlier.

If the associated Weil divisorial sheaf, $\mathscr{O}_X(D)$ is a $\mathbb{Q}$-line bundle, then D is called a **$\mathbb{Q}$-Cartier** divisor. This is equivalent to the property that there exists an $m \in \mathbb{Z}_{>0}$ such that mD is a Cartier divisor.

Let $\mathscr{L}$ be a reflexive $\mathscr{O}_X$-module of rank 1, then since it is a line bundle on the dense open set $X \setminus \operatorname{Sing} X$, similarly to the case of line bundles, there is a rational map $\phi_{\mathscr{L}} : X \dashrightarrow \mathbb{P}(H^0(X, \mathscr{L}))$.

The **Kodaira dimension** of $\mathscr{L}$, a reflexive $\mathscr{O}_X$-module of rank 1 is

$$\kappa(\mathscr{L}) = \max\{\dim \phi_{\mathscr{L}^{[m]}}(X) | m > 0\}.$$

If $H^0(X, \mathscr{L}^{[m]}) = 0$ for all $m > 0$ we let $\kappa(\mathscr{L}) = -1$. Equivalently we have $\kappa(\mathscr{L}) = \operatorname{tr.deg.}_{\mathbb{C}} R(X, \mathscr{L}) - 1$ where $R(X, \mathscr{L}) = \oplus_{m\geq 0} H^0(X, \mathscr{L}^{[m]})$ and that

$$\kappa(\mathscr{L}) = \max\left\{ k \,\middle|\, \limsup \frac{h^0(X, \mathscr{L}^{[m]})}{m^k} > 0 \right\}.$$

We say that $\mathscr{L}$ is **big** if $\kappa(\mathscr{L}) = \dim X$. If $\mathscr{L}$ is a line bundle, this is equivalent to the earlier definition. The Kodaira dimension of a reduced pair (X, D) is defined as the Kodaira dimension of the Weil divisorial sheaf associated to D, i.e., $\kappa(X, D) := \kappa(\mathscr{O}_X(D))$. Similarly, D is **big**, if $\mathscr{O}_X(D)$ is big. In addition, if $D \in \operatorname{Div}(X)$ is nef, then D is big if and only if $D^{\dim X} > 0$.

If X is a smooth variety of dimension n, then the tangent bundle T_X is a vector bundle of rank n and the **canonical line bundle** is defined by $\omega_X = \wedge^n T_X^*$. If X is normal, we let $\omega_X = j_*\omega_{X_{\mathrm{reg}}}$ where $X_{\mathrm{reg}} = X \setminus \operatorname{Sing} X$ is the open subset of smooth points on X and $j : X_{\mathrm{reg}} \to X$ is the corresponding inclusion. It follows that ω_X is a reflexive $\mathscr{O}_X$-module of rank 1 and hence a Weil divisorial sheaf. A **canonical divisor** $K_X \in \operatorname{WDiv}(X)$ is a divisor such that $\mathscr{O}_X(K_X) \simeq \omega_X$. Note that K_X is only defined modulo linear equivalence, nevertheless it is often called *the* canonical divisor.

A smooth projective variety X is of **general type** if ω_X is big, or equivalently if $\kappa(X) = \dim X$. It is easy to see that this condition is invariant under birational equivalence between smooth projective varieties. An arbitrary projective variety is of **general type** if so is any one of its desingularizations. A projective variety is **canonically polarized** if ω_X is ample. Notice that if a smooth projective variety is canonically polarized, then it is of general type. A reduced pair (X, D) is of **log general type** if $\omega_X(D)$ is big, or equivalently $\kappa(X, D) = \dim X$.

2.D Cyclic covers

Let X be a normal variety and $\mathscr{L}$ a $\mathbb{Q}$-line bundle of index m. Assume that $\mathscr{L}^{[m]} \simeq \mathscr{O}_X$ and let $\vartheta : \mathscr{O}_X \to \mathscr{L}^{[m]}$ be a trivialization. Consider the $\mathscr{O}_X$-algebra

$$\mathscr{A} = \bigoplus_{j=0}^{m-1} \mathscr{L}^{[-j]} \simeq \bigoplus_{j=0}^{\infty} \mathscr{L}^{[-j]} t^j \Big/ \left(t^m - \vartheta\right)$$

so that for $i + j \geq m$ multiplication is defined by the rule

$$\mathscr{L}^{[-i]} \otimes \mathscr{L}^{[-j]} \longrightarrow \mathscr{L}^{[-i-j]} \xrightarrow{\ \vartheta\ } \mathscr{L}^{[-i-j]} \otimes \mathscr{L}^{[m]} \xrightarrow{\ \simeq\ } \mathscr{L}^{[m-i-j]}$$

as in [KM98, 2.50]. Let $\sigma : Y := \operatorname{Spec}_X \mathscr{A} \to X$ be the induced finite cover.

EXERCISE 2.5. Prove that $(\sigma^*\mathscr{L})^{**}$, the reflexive pull-back of $\mathscr{L}$, is a line bundle on Y.

If $\mathscr{L} = \omega_X$, then the corresponding $\sigma : Y \to X$ is called the (local) **index** 1 **cover** of X. The previous exercise explains the name and implies that σ^*K_X is a Cartier divisor.

EXERCISE 2.6. Prove that if $\mathscr{L}$ is a line bundle, then $\sigma : Y \to X$ is étale.

EXERCISE 2.7. Formulate the global version of this construction: Let X be a quasi-projective variety, $\mathscr{L}$ a $\mathbb{Q}$-line bundle of index m. If $H^0(X, \mathscr{L}^{[m]}) \neq 0$, then there exists a finite morphism $\sigma : Y \to X$ of degree m such that $(\sigma^*\mathscr{L})^{**}$ is a line bundle.

REMARK 2.8. The problem with the construction in (2.7) is that in this generality one has no control over the singularities of Y. However, if $\mathscr{L}^{[m]}$ is generated by global sections, then it is usually possible to choose σ in such a way that the singularities of Y are not worse than those of X. For more on this, see [KM98, 2.51, 5.21].

2.E $\mathbb{R}$-divisors in the relative setting

If $f : X \to U$ is a proper morphism of normal varieties and $D_i \in \mathrm{WDiv}_{\mathbb{R}}(X)$, then $D_1 \sim_{\mathbb{R},U} D_2$ (i.e. D_1 and D_2 are **$\mathbb{R}$-linearly equivalent over** U) if $D_1 - D_2 = \sum r_i(q_i) + f^*C$ where $r_i \in \mathbb{R}$, q_i are rational functions on X and C is an $\mathbb{R}$-Cartier divisor on U.

Two $\mathbb{R}$-Cartier divisors D_1 and D_2 are **numerically equivalent over** U, $D_1 \equiv_U D_2$ if $(D_1 - D_2) \cdot C = 0$ for any irreducible curve $C \subset X$ which is contained in a fiber.

An $\mathbb{R}$-Cartier divisor D is **nef over** U if $D \cdot C \geq 0$ for any curve $C \subset X$ contracted by f.

An $\mathbb{R}$-Cartier divisor D is ample over U if $D \sim_{\mathbb{R},U} \sum r_i D_i$ where $r_i \in \mathbb{R}_{\geq 0}$ and D_i are ample Cartier divisors over U. An $\mathbb{R}$-Cartier divisor D is **semiample over** U if there is a morphism $g\colon X \to Y$ over U such that $D \sim_{\mathbb{R},U} g^*D'$ where D' is ample over U. An $\mathbb{R}$-Cartier divisor D is **big over** U if

$$\limsup \frac{\mathrm{rk} f_*\mathcal{O}_X(mD)}{m^{\dim f}} > 0$$

where $\dim f$ denotes the dimension of a general fiber of f. Note that D is big over U if and only if $D \sim_{\mathbb{R},U} A + B$ where A is ample over U and $B \geq 0$. An $\mathbb{R}$-Cartier divisor D is **pseudo-effective over** U if its numerical class belongs to the closure of the cone of big divisors over U.

The **real linear system** over U associated to an $\mathbb{R}$-divisor D on X is

$$|D/U|_{\mathbb{R}} = \{D' \geq 0 | D' \sim_{\mathbb{R},U} D\}.$$

Similarly we let $|D/U|_{\mathbb{Q}} = \{D' \geq 0 | D' \sim_{\mathbb{Q},U} D\}$.

The **stable base locus** of D over U is the Zariski closed subset

$$\mathbf{B}(D/U) := \bigcap_{C \in |D/U|_{\mathbb{R}}} \mathrm{Supp}(C).$$

Note that if $|D/U|_{\mathbb{R}} = 0$, then we let $\mathbf{B}(D/U) = X$. When D is $\mathbb{Q}$-Cartier then $\mathbf{B}(D/U)$ is the usual stable base locus.

The **stable fixed divisor** of D over U is $\mathbf{Fix}(D/U)$ the biggest divisor $F \geq 0$ such that $D' \geq F$ for all $D' \in |D/U|_{\mathbb{R}}$.

The **augmented stable base locus** of D over U is given by $\mathbf{B}(D - \varepsilon A/U)$ for any divisor A ample over U and any $0 < \varepsilon \ll 1$. This set is denoted by $\mathbf{B}_+(D/U)$.

The **diminished stable base locus** of D over U is given by $\cup_{\varepsilon>0}\mathbf{B}(D + \varepsilon A/U)$ for any divisor A ample over U. This set is denoted by $\mathbf{B}_-(D/U)$. Note that $\mathbf{B}_-(D/U)$ is a countable union of Zariski closed subsets of X.

Proposition 2.9. *Let $D \geq 0$ be an $\mathbb{R}$-Cartier divisor, then there exists an $\mathbb{R}$-Cartier divisor $D' \geq 0$ such that $D' \sim_{\mathbb{R},U} D$ and if G is an irreducible component of* $\operatorname{Supp}(D')$ *then either mG is mobile for some integer $m > 0$ or G is contained in $\mathbf{B}(D/U)$.*

Proof. We write $D \sim_{\mathbb{R},U} D' = M + F \geq 0$ where every component of F is contained in $\mathbf{B}(D/U)$ and no component of M is contained in $\mathbf{B}(D/U)$. Let $n = n(D')$ be the number of components G of $\operatorname{Supp}(M)$ such that mG is not mobile for any integer $m > 0$. We may choose D' such that $n = n(D')$ is minimal among all divisors $D' \in |D/U|_{\mathbb{R}}$. We proceed by induction on the number n. If $n = 0$ we are done, so assume that $n > 0$ and pick G a component as above. Since G is not contained in $\mathbf{B}(D/U)$, there exists $D'' \in |D/U|_{\mathbb{R}}$ such that G is not contained in the support of D''. Let $E = D' \wedge D''$, then $D' - E \sim_{\mathbb{R},U} D'' - E$ are effective $\mathbb{R}$-divisor whose supports have no common components, and so by (2.10) $D' - E \sim_{\mathbb{R},U} D_1$ where every component of D_1 has a multiple which is mobile. But then $D \sim_{\mathbb{R},U} D_1 + E$ and $n(D_1 + E) < n(D')$. This is a contradiction and the proposition is proven. □

Lemma 2.10. *Let $\pi : X \to U$ be a projective morphism of normal varieties, $D \in \operatorname{WDiv}_{\mathbb{R}}(X)$ and $D_i \in |D/U|_{\mathbb{R}}$ such that $D_1 \wedge D_2 = 0$, then there exists $D' \in |D/U|_{\mathbb{R}}$ such that for any component G of* $\operatorname{Supp}(D')$ *there is an integer $m > 0$ such that mG is mobile.*

Proof. We may write $D_1 - D_2 = \sum r_i(f_i) + \pi^* H$ where $r_i \in \mathbb{R}$, $f_i \in \mathbb{C}(X)$ and $H \in \operatorname{Div}_{\mathbb{R}}(U)$. We may write $H \sim_{\mathbb{R}} -H_1 + H_2$ where the H_i are ample $\mathbb{R}$-divisors on U. Replacing D_i by $D_i + \pi^* H_i$ (and modifying the r_i and the f_i), we may assume that

$$D_1 - D_2 = \sum_{i=1}^{k} r_i(f_i).$$

Assume now that r_i and f_i are chosen so that k is minimal. By (2.13), it follows that $r_1, \ldots, r_k$ are linearly independent over $\mathbb{Q}$. In particular every component of $\operatorname{Supp}(f_i)$ is either a component of D_1 or of D_2. We now choose $q_i \in \mathbb{Q}$ so that $|q_i - r_i| \ll 1$ and we write

$$D'_1 - D'_2 = \sum_{i=1}^{k} q_i(f_i)$$

where $D'_i \geq 0$ and $D'_1 \wedge D'_2 = 0$. It follows that $\operatorname{Supp}(D'_i) = \operatorname{Supp}(D_i)$. Since $D'_i \in \operatorname{WDiv}_{\mathbb{Q}}(X)$, there is a divisor $D^\# \in |D'_1|_{\mathbb{Q}}$ such that every component of $\operatorname{Supp}(D^\#)$ has a multiple which is mobile. We now pick $\lambda > 0$ maximal such that $D_i - \lambda D'_i \geq 0$ for $i \in \{1, 2\}$. We then have that $D_1 - \lambda D'_1 \sim_{\mathbb{R},U} D_2 - \lambda D'_2$ and $(D_1 - \lambda D'_1) \wedge (D_2 - \lambda D'_2) = 0$. Since the number of components of $\operatorname{Supp}(D_1 - \lambda D'_1 + D_2 - \lambda D'_2)$ is strictly less than the number of components of $\operatorname{Supp}(D_1 + D_2)$, we may assume that there is a divisor $D^* \in |D_1 - \lambda D'_1/U|_{\mathbb{R}}$ such that any component of D^* has a multiple which is mobile. But as $D^* + \lambda D^\# \in |D_1/U|_{\mathbb{R}}$, the lemma follows. □

EXERCISE 2.11. Show that if $f : Y \to X$ is a proper birational morphism of normal varieties and $D \in \operatorname{Div}(X)$ is a divisor on X and G is a divisor on Y, then

(2.11.1) If D is big (resp. pseudo-effective) then so is f^*D;

(2.11.2) If D is big (resp. pseudo-effective) and F is effective and its support contains all f-exceptional divisors, then $f_*^{-1}D + F$ is big (resp. pseudo-effective);

(2.11.3) Show that if $G \sim G'$, then $f_*G \sim f_*G'$. Deduce from this that if G is big (resp. pseudo-effective) then so is f_*G;

(2.11.4) Give an example where G is not big, but f_*G is big.

EXERCISE 2.12. Let X be a normal variety and $D_i \in \mathrm{WDiv}_{\mathbb{Q}}(X)$ such that $D_i \geq 0$, $D_1 \sim_{\mathbb{Q}} D_2$ and $D_1 \wedge D_2 = 0$. Show that there is an element $D_3 \in |D_1|_{\mathbb{Q}}$ such that any component of $\mathrm{Supp}(D_3)$ has a multiple which is mobile.

EXERCISE 2.13. Let X be a normal variety and $D \in \mathrm{WDiv}_{\mathbb{R}}(X)$ such that $D = \sum_{i=1}^{k} r_i(f_i)$ where $r_i \in \mathbb{R}$ and $f_i \in \mathbb{C}(X)$. Show that if k is minimal, then $r_1, \ldots, r_k$ are linearly independent over $\mathbb{Q}$.

EXERCISE 2.14. Show that if X is a projective normal variety and G is an $\mathbb{R}$-divisor such that $|G|_{\mathbb{Q}} \neq \emptyset$, then there is an integer $m > 0$ such that $h^0(\mathscr{O}_X(mG)) > 0$. Give an example when $|G|_{\mathbb{R}} \neq \emptyset$ but $|G|_{\mathbb{Q}} = \emptyset$, i.e. $h^0(\mathscr{O}_X(mG)) = 0$ for all $m > 0$.

2.F Families and base change

DEFINITION 2.15. For a morphism $f : Y \to S$ and another morphism $\sigma : T \to S$, the symbol $Y_T = Y_\sigma$ denotes $Y \times_S T$ and $f_T = f_\sigma : Y_T \to T$ the induced natural morphism. In particular, for $t \in S$ we write $Y_t = f^{-1}(t)$. Also, notice that since the setup is symmetric, $T_Y = Y_T$ and there is a natural morphism $\sigma_Y : T_Y \to Y$;

$$\begin{array}{ccc} Y_T = T_Y & \xrightarrow{\sigma_Y} & Y \\ \downarrow{\scriptstyle f_T} & & \downarrow{\scriptstyle f} \\ T & \xrightarrow[\sigma]{} & S \end{array}$$

Similarly, for a sheaf $\mathscr{F}$ on Y, $\mathscr{F}_T = \mathscr{F}_\sigma$ denote by $\sigma_Y^*\mathscr{F}$ the pull-back sheaf on Y_T. In addition, if $T = \operatorname{Spec} F$, then Y_T will also be denoted by Y_F.

DEFINITION 2.16. Let B be a smooth variety over k, and $\Delta \subseteq B$ a closed subset. A **family** over B is variety X together with a flat projective morphism $f : X \to B$ with connected fibers. A family $f : X \to B$ is **isotrivial** if $X_a \simeq X_b$ for any pair of general points $a, b \in B$.

DEFINITION 2.17. Let B be a smooth variety over k, and $\Delta \subseteq B$ a closed subset. Further let $h \in \mathbb{Q}[t]$ be a polynomial. A family $f : X \to B$ is **admissible** (with respect to (B, Δ) and h) if

(2.17.1) X is smooth,

(2.17.2) $f : X \to B$ is not isotrivial,

(2.17.3) Δ contains the discriminant locus of f, i.e., the map $f : X \setminus f^{-1}(\Delta) \to B \setminus \Delta$ is smooth, and

(2.17.4) X_b is projective and ω_{X_b} is ample with Hilbert polynomial $\chi(X, \omega_{X_b}^{\otimes m}) = h(m)$ for all $b \in B \setminus \Delta$.

Two admissible families are **equivalent** if they are isomorphic over $B \setminus \Delta$.

If the relative dimension of f is 1 we will use the genus of X_b instead of the Hilbert polynomial of ω_{X_b}. Recall that in dimension 1 these are equivalent pieces of information.

2.G Parameter spaces and deformations of families

In general, deforming an object means to include that object in a family. We will discuss this at length with respect to moduli spaces.

Given a family $f: X \to B$, we say that B **parameterizes** the members of the family. If all members of a class $\mathfrak{C}$ of varieties appear as fibers of f and all fibers are members of $\mathfrak{C}$, then we say that B is a **parameter space** for the class $\mathfrak{C}$. Note that we do not require that the members of $\mathfrak{C}$ appear only once in the family as with moduli spaces cf. (11.13).

We will see a very useful parameter space; the Hilbert scheme, Chapter 12. For more on parameter spaces in general see [Harr95, Lectures 4, 21].

We will also study something slightly different and there is a potentially confusing point here. Our main objects of study are families, that is, deformations of their members. However, in Chapter 16 we will consider our families as the objects and not the deformations. In other words, we want to look at deformations *of* these families. This works just the same way as deformations of other objects. In addition, we want to fix the base of these families, so we are interested in deformations leaving the base fixed, which makes both the notation and the theory easier.

A **deformation** of a family $f: X \to B$ with the base fixed is a family $\mathscr{X} \to B \times T$, where T is connected and for some $t_0 \in T$ we have $\left(\mathscr{X}_{t_0} \to B \times \{t_0\}\right) \simeq (X \to B)$:

$$\begin{array}{ccc} X \simeq \mathscr{X}_{t_0} & \longrightarrow & \mathscr{X} \\ {\scriptstyle f}\downarrow & & \downarrow \\ B \simeq B \times \{t_0\} & \longrightarrow & B \times T. \end{array}$$

We say that two families $X_1 \to B$ and $X_2 \to B$ have the same **deformation type** if they can be deformed into each other, i.e., if there exists a connected T and a family $\mathscr{X} \to B \times T$ such that for some $t_1, t_2 \in T$, $\left(\mathscr{X}_{t_i} \to B \times \{t_i\}\right) \simeq (X_i \to B)$ for $i = 1, 2$.

We will consider deformations of admissible families. It will be advantageous to restrict to deformations of the family over $B \setminus \Delta$. Doing so potentially allows more deformations than over the original base B: it can easily happen that a deformation over $B \setminus \Delta$, that is, a family $\mathscr{X} \to (B \setminus \Delta) \times T$, cannot be compactified to a (flat) family over $B \times T$, because the compactification could contain fibers of higher than expected dimension. This however, will not cause any problems.

CHAPTER 3

Singularities

In addition to the fact that the mmp forces us to work with singularities, it is also necessary to allow our objects to have singularities in order to accomplish our goal of classifying all canonical models via constructing moduli spaces.

Even if we were only interested in smooth objects, their degenerations provide important information. In other words, it is always useful to find complete moduli problems, i.e., extend our moduli functor so that it admits a complete/compact (and preferably projective) coarse moduli space. This also leads to having to consider singular varieties.

On the other hand, we will have to be careful to limit the kind of singularities that we allow in order to be able to handle them and to obtain a moduli space with a reasonable geometry. It turns out that the constraints imposed by these (and other) issues lead roughly to the same class of singularities.

We will start by considering some examples.

3.A Canonical singularities

For an excellent introduction to this topic the reader is urged to take a thorough look at Miles Reid's *Young person's guide to canonical singularities* [Rei87]. Here we will only briefly touch on the subject.

Let X be a minimal surface of general type that contains a (-2)-curve (a smooth rational curve with self-intersection -2). For an example of such a surface consider the following.

EXAMPLE 3.1. $\widetilde{X} = (x^5+y^5+z^5+w^5=0) \subseteq \mathbb{P}^3$ with the $\mathbb{Z}_2$-action that interchanges $x \leftrightarrow y$ and $z \leftrightarrow w$. This action has five fixed points, $[1:1:-\varepsilon^i:-\varepsilon^i]$ for $i = 1,\dots,5$ where ε is a primitive 5-th root of unity. Consequently the quotient $\widetilde{X}/\mathbb{Z}_2$ has five singular points, each a simple double point of type A_1. Let $X \to \widetilde{X}/\mathbb{Z}_2$ be the minimal resolution of singularities. Then X contains five (-2)-curves, the exceptional divisors over the singularities.

Let us return to the general case, that is, X is a minimal surface of general type that contains a (-2)-curve, $C \subseteq X$. As $C \simeq \mathbb{P}^1$, and X is smooth, the adjunction formula gives us that $K_X \cdot C = 0$. Therefore K_X is not ample.

On the other hand, since X is a minimal surface of general type, it follows that K_X is semiample, that is, some multiple of it is base-point free. In other words, there exists a morphism,

$$\phi_m : X \to X_{\mathrm{can}} \subseteq \mathbb{P}(H^0(X, \mathcal{O}_X(mK_X))).$$

This follows from various results, for example Bombieri's classification of pluri-canonical maps, but perhaps the simplest proof is provided by Miles Reid [Rei97, E.3].

It is then relatively easy to see that this morphism onto its image is independent of m. This constant image is called the **canonical model** of X, let us denote it by X_{can}. Note that $X_{\mathrm{can}} = \operatorname{Proj} \oplus_{m\geq 0} H^0(X, \mathcal{O}_X(mK_X))$.

The good news is that the canonical divisor of X_{can} is Cartier and ample, but the trouble is that the surface X_{can} can be singular. However, the singularities that may occur are not too bad, so that we still have a good chance to study the geometry of X via its canonical model X_{can}. In fact, the singularities that can occur on the canonical model of a surface of general type belong to a much studied class. This class goes by several names; they are called **Du Val** singularities, or **rational double points**, or **Gorenstein, canonical**. For more on these singularities, we refer the reader to [Dur79], [Rei87]. It turns out that if $f : Y \to X$ is the minimal resolution of a surface X with canonical singularities, then $K_Y = f^*K_X$. Note also that if $f' : Y' \to X$ is any resolution of X, then f' factors through $\nu : Y' \to Y$ and so $K_{Y'} = \nu^*K_Y + E = f'^*K_X + E$ where E is an effective divisor whose support coincides with $\operatorname{Ex}(\nu)$.

In higher dimensions, the situation is similar. The class of canonical singularities is defined to be exactly the class of singularities occuring in the canonical model $X_{\mathrm{can}} = \operatorname{Proj} \oplus_{m\geq 0} H^0(X, \mathcal{O}_X(mK_X))$ of a projective variety of general type. (Note that X_{can} will exist if and only if $\oplus_{m\geq 0} H^0(X, \mathcal{O}_X(mK_X))$ is finitely generated cf. (5.75).) It turns out that a variety X has canonical singularities if and only if there is a positive integer r such that rK_X is Cartier and for any resolution $f : Y \to X$, we have that $rK_Y = f^*(rK_X) + E$ where E is an effective (and f-exceptional) divisor (cf. [Rei87]). Note that then $H^0(Y, \mathcal{O}_Y(mK_Y)) \simeq H^0(X, \mathcal{O}_X(mK_X))$ for all $m > 0$.

Terminal singularities are similarly defined by requiring that the support of E contain all f-exceptional divisors.

Note that even though in dimension ≥ 3 there is no "minimal desingularization", given a variety X with canonical singularities, we may find a "partial desingularization" $f : Y \to X$ such that Y has terminal singularities and $K_Y = f^*K_X$. It then follows that for any desingularization $f' : Y' \to X$ that factors through a morphism $\nu : Y' \to Y$, we have $K_{Y'} = \nu^* K_Y + E = f'^* K_X + E$ where E is an effective divisor whose support coincides with the set of ν-exceptional divisors cf. (5.41).

3.B Cones

Let $C \subseteq \mathbb{P}^2$ be a curve of degree d and $X \subseteq \mathbb{P}^3$ the projectivized cone over C. As X is a degree d hypersurface, it admits a smoothing.

EXERCISE 3.2. Let $\Xi = (x^d + y^d + z^d + tw^d = 0) \subseteq \mathbb{P}^3_{x:y:z:w} \times \mathbb{A}^1_t$. The special fiber Ξ_0 is a cone over a smooth plane curve of degree d and the general fiber Ξ_t, for $t \neq 0$, is a smooth surface of degree d in $\mathbb{P}^3$.

This, again, suggests that we should deal with some singularities. The question is, whether we can limit the type of singularities we must deal with. In particular, in the above example, can we limit the type of cones we need to deal with?

First we need an auxiliary computation.

EXERCISE 3.3. Let W be a smooth variety and $X = X_1 \cup X_2 \subseteq W$ such that X_1 and X_2 are Cartier divisors in W. Prove that

$$K_X \text{ is ample} \quad \Leftrightarrow \quad K_X\big|_{X_i} = K_{X_i} + X_{3-i}\big|_{X_i} \text{ is ample for } i = 1, 2. \tag{3.3.1}$$

Next, let X be a normal projective surface with K_X ample and an isolated singular point $P \in \operatorname{Sing} X$. Assume that X is isomorphic to a cone $\Xi_0 \subseteq \mathbb{P}^3$ as in Example 3.2 locally analytically near P. Further assume that X is the special fiber of a smoothing family Ξ that itself is smooth. We would like to see whether we may resolve the singular point $P \in X$ and still stay within our moduli problem, i.e., that K would remain ample. For this purpose we may assume that P is the only singular point of X.

Let $\Upsilon \to \Xi$ be the blowing-up of $P \in \Xi$ and let $\widetilde{X}$ denote the strict transform of X. Then $\Upsilon_0 = \widetilde{X} \cup E$ where $E \simeq \mathbb{P}^2$ is the exceptional divisor of the blow-up. Clearly, $\sigma : \widetilde{X} \to X$ is the blow-up of P on X, so it is a smooth surface and $\widetilde{X} \cap E$ is isomorphic to the degree d curve over which X is locally analytically a cone.

We would like to determine the condition on d that ensures that the canonical divisor of Υ_0 is still ample. According to (3.3.1) this means that we need that $K_E + \widetilde{X}\big|_E$ and $K_{\widetilde{X}} + E\big|_{\widetilde{X}}$ be ample.

As $E \simeq \mathbb{P}^2$, $\omega_E \simeq \mathscr{O}_{\mathbb{P}^2}(-3)$, so $\mathscr{O}_E(K_E + \widetilde{X}\big|_E) \simeq \mathscr{O}_{\mathbb{P}^2}(d-3)$. This is ample if and only if $d > 3$.

As this computation is local near P the only relevant issue about the ampleness of $K_{\widetilde{X}} + E\big|_{\widetilde{X}}$ is whether it is ample in a neighbourhood of $E_X := E\big|_{\widetilde{X}}$. By (3.4) this is equivalent to asking when $(K_{\widetilde{X}} + E_X) \cdot E_X$ is positive.

EXERCISE 3.4. Let Z be a smooth projective surface with non-negative Kodaira dimension and $\Gamma \subset Z$ an effective divisor. If $(K_Z + \Gamma) \cdot C > 0$ for every proper curve $C \subset Z$, then $K_Z + \Gamma$ is ample.

Now, observe that by the adjunction formula $(K_{\widetilde{X}} + E_X) \cdot E_X = \deg K_{E_X} = d(d-3)$ as E_X is isomorphic to a plane curve of degree d. Again, we obtain the same condition as above and thus conclude that K_{Υ_0} is ample if and only if $d > 3$.

Since the objects that we consider in our moduli problem must have an ample canonical class, we may only replace X by Υ_0 if $d > 3$. For our moduli problem this means that we have to allow cone singularities over curves of degree $d \leq 3$. The singularity we obtain for $d = 2$ is a rational double point, but the singularity for $d = 3$ is new and it is not even rational.

EXERCISE 3.5. Let $X \subset \mathbb{P}^n$ be a smooth variety and $C_X \subset \mathbb{P}^{n+1}$ be the projectivized cone. For any divisor D on X, consider the divisor corresponding to $C_D \subset C_X$. Show that C_D is $\mathbb{Q}$-Cartier if and only if $\mathscr{O}_X(kD) \simeq \mathscr{O}_X(j)$ for some $k \in \mathbb{Z}_{>0}$ and $j \in \mathbb{Z}$.

3.C Log canonical singularities

Let us investigate the previous situation under more general assumptions.

COMMENTARY 3.6. One of our ultimate goals is to construct a moduli space for canonical models of varieties. We are already aware that the minimal model program has to deal with singularities and so we must allow some singularities on canonical models. We would also like to understand what constraints the goal of constructing a moduli space imposes. The point is that in order to construct our moduli space, the objects must have an ample canonical class. It is possible that a family of canonical models degenerates to a singular fiber that has singularities worse than the original canonical models. An important question then is whether we may resolve the singularities of this special fiber and retain ampleness of the canonical class. The next example shows that this is not always the case.

COMPUTATION 3.7. Let $D = \sum_{i=0}^{r} \lambda_i D_i$ $(\lambda_i \in \mathbb{N})$ be a divisor with only normal crossing singularities in some smooth ambient variety such that $\lambda_0 = 1$. Using a generalized version of the adjunction formula shows that in this situation (3.3.1) remains true.

(3.7.1) $$K_D\big|_{D_0} = K_{D_0} + \sum_{i=1}^{r} \lambda_i D_i\big|_{D_0}.$$

Let $f : \Xi \to B$ a projective family with $\dim B = 1$, Ξ smooth and K_{Ξ_b} ample for all $b \in B$. Further let $X = \Xi_{b_0}$ for some $b_0 \in B$ a singular fiber and let $\sigma : \Upsilon \to \Xi$ be an embedded resolution of $X \subseteq \Xi$. Finally let $Y = \sigma^* X = \widetilde{X} + \sum_{i=1}^{r} \lambda_i F_i$ where $\widetilde{X}$ is the strict transform of X and F_i are exceptional divisors for σ. We are interested in finding conditions that are necessary for K_Y to remain ample (cf. (3.6)).

Let $E_i := F_i\big|_{\widetilde{X}}$ be the exceptional divisors for $\sigma : \widetilde{X} \to X$ and for simplicity of computation, assume that the E_i are irreducible. For K_Y to be ample we need that $K_Y\big|_{\widetilde{X}}$ as well as $K_Y\big|_{F_i}$ for all i are all ample. We focus on $K_Y\big|_{\widetilde{X}}$ for which by (3.7.1) we have that

$$K_Y|_{\widetilde{X}} = K_{\widetilde{X}} + \sum_{i=1}^{r} \lambda_i E_i.$$

As usual, we may write $K_{\widetilde{X}} = \sigma^* K_X + \sum_{i=1}^{r} a_i E_i$, so we are looking for conditions to guarantee that $\sigma^* K_X + \sum(a_i + \lambda_i)E_i$ is ample. In particular, its restriction to any of the E_i has to be ample. To further simplify our computation let us assume that $\dim X = 2$. Then the condition that we want satisfied is that for all j,

$$\left(\sum_{i=1}^{r}(a_i + \lambda_i)E_i\right) \cdot E_j > 0. \tag{3.7.2}$$

Let

$$E_+ = \sum_{a_i+\lambda_i \geq 0} |a_i + \lambda_i| E_i, \quad \text{and}$$

$$E_- = \sum_{a_i+\lambda_i < 0} |a_i + \lambda_i| E_i, \quad \text{so}$$

$$\sum_{i=1}^{r}(a_i + \lambda_i)E_i = E_+ - E_-.$$

Choose a j such that $E_j \subseteq \operatorname{supp} E_+$. Then $E_- \cdot E_j \geq 0$ since $E_j \nsubseteq E_-$ and since (3.7.2) implies that $(E_+ - E_-) \cdot E_j > 0$. These together imply that $E_+ \cdot E_j > 0$ and so $E_+^2 > 0$. However, the E_i are exceptional divisors of a birational morphism, so their intersection matrix, $(E_i \cdot E_j)$ is negative definite.

The only way this can happen is if $E_+ = 0$. In other words, $a_i + \lambda_i < 0$ for all i. However, the λ_i are positive integers, so a necessary condition for K_Y to remain ample is that $a_i < -1$ for all $i = 1, \dots, r$.

The definition of a *log canonical* singularity is the exact opposite of this condition. It requires that X be normal and for any resolution of singularities, say $Y \to X$, we have that $a_i \geq -1$ for all i. The above argument shows that we stand a fighting chance to preserve the ampleness of the canonical bundle if we resolve singularities that are *worse* than log canonical, but have no hope to do so with log canonical singularities. In other words, this is another class of singularities that we have to allow in order to construct moduli spaces of canonically polarized varieties. Actually, the class of singularities we obtained for the cones in the previous subsection belong to this class as well. In fact, all the normal singularities that we have considered so far belong to this class.

The good news is that by now we have covered pretty much all the ways that something can go wrong and found the class of normal singularities we must allow for our moduli problem.

It turns out that we also have to deal with some non-normal singularities and in fact in the above example we have not really needed that X be normal. In the next few subsections we will see examples of non-normal singularities that we will have to handle. In particular, we will see that we have to allow the non-normal cousins of log canonical singularities. These are called *semi-log canonical* singularities and the reader can find their definition in Section 3.F.

3.D Normal crossings

A **normal crossing** singularity is one that is locally analytically (or formally) isomorphic to the intersection of coordinate hyperplanes in a linear space. In other words, it is a singularity locally analytically defined as $(x_1x_2\cdots x_r = 0) \subseteq \mathbb{A}^n$ for some $r \leq n$. In particular, as opposed to the curve case, for surfaces it allows for triple intersections. However, triple (or higher) intersections may be "semi-resolved": Let $X = (xyz = 0) \subseteq \mathbb{A}^3$. Blow-up the origin $O \in \mathbb{A}^3$, $\sigma : \mathrm{Bl}_O\mathbb{A}^3 \to \mathbb{A}^3$ and consider the strict transform of X, $\sigma : \widetilde{X} \to X$. Observe that $\widetilde{X}$ has only double normal crossings and the morphism σ is an isomorphism over $X \setminus \{O\}$. Therefore, this is a semi-resolution as defined in (3.13.4). Double normal crossings cannot be resolved the same way, because the double locus is of codimension 1, so any morphism from any space with any kind of singularities that are not double normal crossings would fail to be an isomorphism in codimension 1.

Since normal crossings are (analytically) locally defined by a single equation, they are Gorenstein and hence the canonical sheaf ω_X is still a line bundle and so it makes sense to require it to be ample.

These singularities naturally appear in the construction of moduli spaces. Consider for example the moduli space of stable curves which is a natural compactification of the moduli space of smooth projective curves. A stable curve is a projective curve with only normal crossing singularities and an ample canonical bundle. These are naturally degenerations of smooth projective curves of the same genus. By stable reduction (cf. [KKMSD73], [HMo98, §3.C], [KM98, §7.2]) every degeneration may be "resolved" to a stable one. These together imply that the moduli functor of stable curves of a given genus is a natural compactification of the moduli functor of smooth curves of the same genus.

As we want to understand degenerations of our preferred families, we have to allow normal crossings.

Another important point to remember about normal crossings is that they are *not* normal. In particular they do not belong to the previous category. For some interesting and perhaps surprising examples of surfaces with normal crossings see [Kol07c].

3.E Pinch points

Another non-normal singularity that can occur as the limit of smooth varieties is a **pinch point** singularity. It is locally analytically defined as $(x_1^2 = x_2x_3^2) \subseteq \mathbb{A}^n$ $(n \geq 3)$. This singularity is a double normal crossing away from the pinch point. Its normalization is smooth, but blowing up the pinch point does not make it any better as shown by the example that follows.

EXAMPLE 3.8. Let $X = (x_1^2 = x_2^2x_3) \subseteq \mathbb{A}^3$, where x_1, x_2, x_3 are linear coordinates on $\mathbb{A}^3$, $O = (0,0,0)$ and compute $\mathrm{Bl}_O X$. First, recall that

$$\mathrm{Bl}_O\mathbb{A}^3 = (x_i y_j = x_j y_i | i, j = 1, 2, 3) \subset \mathbb{A}^3 \times \mathbb{P}^2,$$

where y_1, y_2, y_3 are homogenous coordinates on $\mathbb{P}^2$.

(3.8.1) Assume that $y_1 = 1$. Then $x_2 = x_1 y_2$ and $x_3 = x_1 y_3$ and the equation of the preimage of X becomes $x_1^2 = x_1^3 y_2^2 y_3$. This breaks up into $x_1^2 = 0$ and $1 = x_1 y_2^2 y_3$. The former equation defines the exceptional divisor and the latter defines the strict transform of X, i.e., $\mathrm{Bl}_O X$. This does not have any points over $O \in X$, so on this chart, the blow-up morphism $\mathrm{Bl}_O X \to X$ is an isomorphism and $\mathrm{Bl}_O X$ is smooth.

(3.8.2) Assume that $y_2 = 1$. Then $x_1 = x_2 y_1$ and $x_3 = x_2 y_3$ and the equation of the preimage of X becomes $x_2^2 y_1^2 = x_2^3 y_3$. This breaks up into $x_2^2 = 0$ and $y_1^2 = x_2 y_3$. Again, the former equation defines the exceptional divisor and the latter the strict transform of X, $\mathrm{Bl}_O X$. Notice that on this chart a coordinate system is given by x_2, y_1, y_3 and the equation defines a quadric cone cf. Section 3.B. Then blowing up the vertex of the cone gives a resolution on this chart.

(3.8.3) Assume that $y_3 = 1$. Then $x_1 = x_3 y_1$ and $x_2 = x_3 y_2$ and the equation of the preimage of X becomes $x_3^2 y_1^2 = x_3^3 y_2^2$. This breaks up as $x_3^2 = 0$ and $y_1^2 = y_2^2 x_3$. Again, the former equation defines the exceptional divisor and the latter the strict transform of X, $\mathrm{Bl}_O X$. Notice that on this chart a coordinate system is given by x_3, y_1, y_3 and the latter equation is the same as the one we started with. So, $\mathrm{Bl}_O X$ again has a pinch point.

This computation shows that the blow-up of a pinch point will be, if anything, more singular than the original and at best it can be resolved to be a pinch point again.

From this example we conclude that a pinch point singularity cannot be resolved or even just made somewhat "better" by only trying to change it over the pinch point. It may only be resolved by taking the normalization. As in the case of double normal crossings, this is not an isomorphism in codimension 1.

OBSERVATION 3.9. Double normal crossings and pinch points share the following interesting properties:

(3.9.1) Their normalization is smooth.

(3.9.2) The normalization morphism is *not* an isomorphism in codimension 1.

(3.9.3) It is not possible to find a partial resolution that is an isomorphism in codimension 1 that would make them better in any reasonable sense.

REMARK 3.10. Notice that all normal crossings share the first two properties, but, in dimension at least 2, not the third one as they may be partially resolved to double normal crossings.

We conclude that double normal crossing and pintch point singularities are unavoidable. However, at the same time, they should be viewed as the simplest non-normal singularities. In fact, in some sense they are much simpler than most normal singularities.

Furthermore, all other singularities can be resolved to these: Any reduced scheme admits a partial resolution to a scheme with only double normal crossings and pinch points such that the resolution morphism is an isomorphism wherever the original scheme is smooth, or has only double normal crossings or pinch points [Kol08c]. Of course, this only gives a partial resolution that is an isomorphism in codimension 1 if the scheme we start with has double normal crossings in codimension 1 already. However, this turns out to be a condition we can achieve.

We will discuss relevant partial resolutions in more detail in (3.13).

3.F Semi-log canonical singularities

As a warm-up, let us first define the normal and more traditional singularities that are relevant in the minimal model program.

DEFINITION 3.11. A normal variety X is called **$\mathbb{Q}$-Gorenstein** if K_X is $\mathbb{Q}$-Cartier, i.e., some integer multiple of K_X is a Cartier divisor. For a $\mathbb{Q}$-Gorenstein variety X, the **index** of X is the smallest positive integer m such that mK_X is Cartier cf. Section 2.C.

Let X be a $\mathbb{Q}$-Gorenstein variety and $f : \widetilde{X} \to X$ a resolution of singularities with normal crossings exceptional divisor $E = \cup E_i$. Express the canonical divisor of $\widetilde{X}$ in terms of K_X and the exceptional divisors:

$$K_{\widetilde{X}} = f^*K_X + \sum a_i E_i$$

where $a_i \in \mathbb{Q}$ and $f_*K_{\widetilde{X}} = K_X$. Then

$$X \text{ has } \begin{array}{c}\textbf{terminal}\\ \textbf{canonical}\\ \textbf{log terminal}\\ \textbf{log canonical}\end{array} \text{ singularities if } \begin{array}{r} a_i > 0,\\ a_i \geq 0,\\ a_i > -1,\\ a_i \geq -1,\end{array}$$

for all i and any resolution f as above.

EXERCISE 3.12. Let X be a variety with only canonical index 1 singularities. Prove that then for any resolution of singularities $f : \widetilde{X} \to X$,

(3.12.1) $f^*\omega_X \subseteq \omega_{\widetilde{X}}$, and

(3.12.2) $f_*\omega_{\widetilde{X}} \simeq \omega_X$.

DEFINITION 3.13. Let X be a scheme of dimension n and $x \in X$ a closed point.

(3.13.1) $x \in X$ is a **double normal crossing** if it is locally analytically (or formally) isomorphic to the singularity

$$\{0 \in (x_0x_1 = 0)\} \subseteq \left\{0 \in \mathbb{A}^{n+1}\right\},$$

where $n \geq 1$.

(3.13.2) $x \in X$ is a **pinch point** if it is locally analytically (or formally) isomorphic to the singularity

$$\left\{0 \in (x_0^2 = x_1^2x_2)\right\} \subseteq \left\{0 \in \mathbb{A}^{n+1}\right\},$$

where $n \geq 2$.

(3.13.3) X is **semi-smooth** if all closed points of X are either smooth, or a double normal crossing, or a pinch point. In this case, unless X is smooth, $D_X := \operatorname{Sing} X \subseteq X$ is a smooth $(n-1)$-fold. If $\nu : \widetilde{X} \to X$ is the normalization, then $\widetilde{X}$ is smooth and $\widetilde{D}_X := \nu^{-1}(D_X) \to D_X$ is a double cover ramified along the pinch locus.

(3.13.4) A morphism, $f : Y \to X$ is a **semi-resolution** if

- f is proper,
- Y is semi-smooth,
- no component of D_Y is f-exceptional, and
- there exists a closed subset $Z \subseteq X$, with $\operatorname{codim}(Z, X) \geq 2$ such that

$$f\big|_{f^{-1}(X\setminus Z)} : f^{-1}(X \setminus Z) \xrightarrow{\simeq} X \setminus Z$$

is an isomorphism.

Let E denote the exceptional divisor (i.e., the codimension 1 part of the exceptional set, not necessarily the whole exceptional set) of f. Then f is a **good semi-resolution** if $E \cup D_Y$ is a divisor with global normal crossings on Y.

(3.13.5) X has **semi-log canonical** (resp. **semi-log terminal**) singularities if

(a) X is reduced,

(b) X is S_2,

(c) K_X is $\mathbb{Q}$-Cartier, and

(d) there exists a good semi-resolution of singularities $f : \widetilde{X} \to X$ with exceptional divisor $E = \cup E_i$, and we write $K_{\widetilde{X}} \equiv f^* K_X + \sum a_i E_i$ with $a_i \in \mathbb{Q}$, then $a_i \geq -1$ (resp. $a_i > -1$) for all i.

REMARK 3.14. Note that a semi-smooth scheme has at worst hypersurface singularities, so in particular it is Gorenstein (see (3.53)). This implies that a semi-log canonical variety is Gorenstein in codimension 1.

DEFINITION 3.15. A variety that is Gorenstein in codimension 1 is called G_1.

REMARK 3.16. Recall that being normal is equivalent to R_1 and S_2, that is, X is normal if and only if it is non-singular in codimension 1 and satisfies Serre's S_2 condition. At first sight this condition may suggest that the best thing about being normal is that it is non-singular in codimension 1. It turns out that in some respect the S_2 condition is much more important. At the same time, it is harder to have a geometric intuition about this condition. The main geometric meaning of the S_2 condition is that on an S_2 variety, functions defined in codimension 1, i.e., outside a codimension 2 subvariety, can be extended to the entire variety cf. (3.17).

This is what makes normal varieties so useful: by the R_1 property there exists a subvariety of codimension at least 2 that contains all the singularities, and then the S_2 property allows one to work with functions (including sections of line bundles) as if they were defined on a non-singular variety.

Here we have to make do without normality. However, observe that despite the very technical definition of a semi-log canonical variety, it has some useful properties. For one it is S_2, which is of major importance and although it is not R_1, it is at least G_1, that is, it is Gorenstein in codimension 1 cf. (3.14), (3.15).

To get a feeling for the S_2 condition, we suggest working on the following exercise:

EXERCISE 3.17. Let X be a variety and $x \in X$ a point. Prove that then depth $\mathscr{O}_{X,x} \geq 2$ if and only if for any open neighborhood U of x, every section of $\mathscr{O}_X$ over $U \setminus \{x\}$ extends uniquely to a section over U. In other words, if $\iota : X \setminus \{x\} \hookrightarrow X$ is the embedding, then depth $\mathscr{O}_{X,x} \geq 2$ if and only if $\iota_* \mathscr{O}_{X \setminus \{x\}} \simeq \mathscr{O}_X$.

REMARK 3.18. It is relatively easy to prove that if X has semi-log canonical (resp. semi-log terminal) singularities, then the condition in (3.13.5d) follows for *all* good semi-resolutions.

REMARK 3.19. One may further generalize the notion of semi-log canonical and define **weakly semi-log canonical** singularities as those that are seminormal, S_2 and with an appropriately chosen divisor on the normalization, that pair is log canonical. In this context semi-log canonical singularities are exactly those weakly semi-log canonical divisors that are Gorenstein in codimension 1. For the precise definition and more details on these singularities and their relationships see [KSS10].

REMARK 3.20. In the definition of a semi-resolution, one could choose to require that the exceptional set be a divisor. This leads to slightly different notions and at the time of the writing of this article it has not been settled whether either of the definitions and the notions of singularities they lead to are unnecessary. For more on singularities related to semi-resolutions see [KSB88], [Kol92], and [Kol08c].

3.G Pairs

If X is a normal variety and $\Delta \in \mathrm{WDiv}_{\mathbb{R}}(X)$ is an effective $\mathbb{R}$-divisor such that $K_X + \Delta \in \mathrm{Div}_{\mathbb{R}}(X)$, then we say that (X, Δ) is a **pair**. A pair (X, Δ) is called a **reduced pair** if all the irreducible components of Δ have multiplicity 1 in Δ, i.e., all the irreducible components of Δ are reduced.

A pair (X, Δ) has **simple normal crossings** (or **snc** for short) if X is smooth, each irreducible component of $\mathrm{Supp}(\Delta)$ is smooth and, locally analytically, $\mathrm{Supp}(\Delta) \subset X$ is isomorphic to the intersection of coordinate hyperplanes in affine space $\{x_1 x_2 \cdots x_r = 0\} \subset \mathbb{A}^n$. Often, we will simply say that Δ has simple normal crossings. In the literature, the term snc is sometimes replaced by log smooth.

A **log resolution** of a pair (X, Δ) is a proper birational morphism $f : Y \to X$ such that $\mathrm{Ex}(f)$ is a divisor and $(Y, f_*^{-1}\Delta + \mathrm{Ex}(f))$ has simple normal crossings. Here $\mathrm{Ex}(f)$ is the **exceptional** set of f i.e. the set of points on Y at which f is not an isomorphism.

REMARK 3.21. For a particularly accessible treatment of resolutions of singularities, we recommend [Kol07b].

We will write

$$K_Y + \Gamma = f^*(K_X + \Delta) + E \qquad (3.21.1)$$

where $f_*K_Y = K_X$, Γ and E are effective $\mathbb{R}$-divisors with no common components, $f_*\Gamma = \Delta$ and $f_*E = 0$. Note that such E and Γ are uniquely determined by $f : Y \to X$. Let F be any prime divisor on Y, then the **discrepancy** of F with respect to (X, Δ) is given by $a(F; X, \Delta) = \operatorname{mult}_F(E - \Gamma)$. Note that for any divisor F over Y, the discrepancy $a(F; X, \Delta)$ is independent of the choice of Y. The **total discrepancy** and the **discrepancy** of a pair (X, Δ) are defined by

$$\operatorname{totaldiscrep}(X, \Delta) := \inf_E\{a(E; X, \Delta) | E \text{ is a divisor over } X\}, \text{ and}$$

$$\operatorname{discrep}(X, \Delta) := \inf_E\{a(E; X, \Delta) | E \text{ is an exceptional divisor over } X\}.$$

If $V \neq \emptyset$ is a linear series on X and f is a log resolution of a log pair (X, Δ) such that $\operatorname{Mov}(f^*V)$ is base point free and $\operatorname{Fix}(f^*V) + f_*^{-1}\Delta + \operatorname{Ex}(f)$ has simple normal crossings, then f is a **log resolution of** (X, Δ) **and** V. If f is a log resolution of $(X, 0)$ and V, we will simply say that f is a log resolution of V.

A **non-kawamata log terminal place** or **non-klt place** of a pair (X, Δ) is a divisor F over X with $a(F; X, \Delta) \leq -1$ and a **center of non-kawamata log terminal singularities** or **non-klt center** is the image of any non-klt place. In the literature, non-klt places and centers are often called log canonical places and centers.

A pair (X, Δ) is **canonical** (resp. **terminal**) if for any divisor F exceptional over X we have that $a(F; X, \Delta) \geq 0$ (resp. $a(F; X, \Delta) > 0$) i.e. if $\operatorname{discrep}(X, \Delta) \geq 0$ (resp. $\operatorname{discrep}(X, \Delta) > 0$). It is known that if (X, Δ) is a canonical surface, then it has at worst Du Val singularities (cf. [KM98, 4.2]) and if (X, Δ) is a terminal surface, then it is smooth.

A pair (X, Δ) is **log canonical** or **lc** (resp. **kawamata log terminal** or **klt**) if for any divisor F over X we have that $a(F; X, \Delta) \geq -1$ (resp. $a(F; X, \Delta) > -1$), i.e., if $\operatorname{totaldiscrep}(X, \Delta) \geq -1$ (resp. $\operatorname{totaldiscrep}(X, \Delta) > -1$). It suffices to check the above condition on any given log resolution of (X, Δ). If $(X, 0)$ is klt, then we say that X is **log terminal**.

A pair (X, Δ) is **divisorially log terminal** or **dlt** if the coefficients of Δ are ≤ 1 and there is a log resolution $f : Y \to X$ of (X, Δ) such that $a(F; X, \Delta) > -1$ for all f-exceptional divisors F on Y. This is equivalent to requiring that there is a closed subset $Z \subset X$ such that $(X - Z, \Delta|_{X-Z})$ has simple normal crossings and, if F is an irreducible divisor over X with center contained in Z, then $a(F; X, \Delta) > -1$ cf. [Sza94]. If S is an irreducible component of the intersection of components of $\lfloor \Delta \rfloor$, then S is normal and the pair (S, Δ_S) defined by adjunction $(K_X + \Delta)|_S = K_S + \Delta_S$ is also dlt. Moreover, we have the following:

Theorem 3.22. *Let (X, Δ) be a dlt pair and V be a linear series whose base locus contains no non-klt centers of (X, Δ). Then there exists a log resolution of (X, Δ) and V which is an isomorphism along the generic point of each stratum of $(X, \lfloor \Delta \rfloor)$.*

Proof. See [Sza94]. □

If (X, Δ) is a pair such that $a(F; X, \Delta) > -1$ for all irreducible f-exceptional divisors F over X, then we say that (X, Δ) is **purely log terminal** or **plt**. In this case $\lfloor \Delta \rfloor = S$ is a disjoint union of normal prime divisors. If S is irreducible, then the pair (S, Δ_S) defined by adjunction $(K_X + \Delta)|_S = K_S + \Delta_S$ is klt.

REMARK 3.23. We will say that $K_X + \Delta$ is lc, klt, dlt, plt, canonical or terminal if so is (X, Δ).

Theorem 3.24 (Adjunction). *Let (X, Δ) be a lc pair and $S \subset \lfloor \Delta \rfloor$ a normal component of $\lfloor \Delta \rfloor$ of coefficient 1. Then there is a naturally defined $\mathbb{R}$-divisor Δ_S on S such that:*

(3.24.1) $(K_X + \Delta)|_S = K_S + \Delta_S$;

(3.24.2) If (X, Δ) is dlt, then S is normal and (S, Δ_S) is dlt;

(3.24.3) If (X, Δ) is plt, then $K_S + \Delta_S$ is klt;

(3.24.4) If (X, S) is plt, then the coefficients of Δ_S at a codimension 1 point P have the form $(r-1)/r$ where r is the index of S at P (equivalently, r is the index of $K_X + S$ at P or r is the order of the cyclic group $\mathrm{Weil}(\mathcal{O}_{X,P})$). It follows that if $\Delta = \sum \delta_i \Delta_i$, then $\mathrm{mult}_P(\Delta|_S) = \sum \frac{m_i \delta_i}{r}$ where $m_i \in \mathbb{Z}_{\geq 0}$;

(3.24.5) If (X, Δ) is plt, $f : Y \to X$ is a proper birational morphism of normal varieties $T = f_*^{-1} S$ and $K_Y + \Gamma = f^*(K_X + \Delta) + E$ as in (3.21.1), then $(f|_T)_*(E|_T) = 0$ and $(f|_T)_*(\Gamma_T) = \Delta_S$ where $(K_Y + \Gamma)|_T = K_T + \Gamma_T$.

Proof. See §16 of [Kol92]. □

REMARK 3.25. In the literature, the divisor Δ_S is known as the *different* and it is often denoted by $\mathrm{Diff}(\Delta - S)$.

REMARK 3.26. Du Val singularities are discussed in §4 of [KM98]. Terminal and canonical 3-fold singularities are discussed in [Rei87]. A generalization of klt, lc, canonical, terminal pairs to the context where $K_X + \Delta$ is not $\mathbb{R}$-Cartier is defined in [HdF08].

EXERCISE 3.27. Let X be the cone over a rational curve of degree n with vertex $O \in X$. Show that $\mathrm{discrep}(X, O) = -1 + 2/n$.

EXERCISE 3.28. Let X be the cone over an elliptic curve. Show that X is lc but not klt. Show that X does not have rational singularities.

EXERCISE 3.29. Let $S \subset X$ be a line on the cone over a rational curve of degree n. Show that (X, S) is plt and compute $(K_X + S)|_S$.

EXERCISE 3.30. Let $S \subset X = \{y^3 = x^2\} \subset \mathbb{C}^2$ be a cusp in the plane. Show that (X, tS) is klt for $t < 5/6$.

EXERCISE 3.31. Show that if (X, Δ) is a pair and $a(F; X, \Delta) < -1$ for some divisor F over X, then $\min\{a(F; X, \Delta)\} = -\infty$.

EXERCISE 3.32. Show that if (X, Δ) is a pair, then either $\mathrm{discrep}(X, \Delta) = -\infty$ or $-1 \leq \mathrm{totaldiscrep.}(X, \Delta) \leq \mathrm{discrep}(X, \Delta) \leq 1$.

EXERCISE 3.33. Show that if $(X, \Delta = \sum d_i \Delta_i)$ is a pair with $0 \leq d_i \leq 1$ and $\mathrm{Supp}(\Delta)$ has simple normal crossings, then $\mathrm{discrep}(X, \Delta)$ is given by the minimum between 1, $1 - d_i$ and $1 - d_i - d_j$ where Δ_i intersects Δ_j.

EXERCISE 3.34. Show that if (X, Δ) is a dlt pair (resp. and $\mathbf{B}(\Delta)$ contains no non-klt center of (X, Δ)), then there exists an $\mathbb{R}$-divisor Δ_0 (resp. $\Delta_0 \sim_{\mathbb{R}} \Delta$) such that (X, Δ_0) is a klt pair. In particular, the non-klt centers of (X, Δ) are contained in $\mathrm{Supp}(\Delta)$.

EXERCISE 3.35. Show that if (X, Δ) klt and $K_X + \Delta$ is big, then for any $0 < \varepsilon \ll 1$, there is a klt pair (X, Δ') such that Δ' is big and $K_X + \Delta' \sim_{\mathbb{R}} (1 + \varepsilon)(K_X + \Delta)$.

EXERCISE 3.36. Let (X, Δ) be a klt pair, then we may find a projective birational morphism $f : Y \to X$ such that if we write $K_Y + \Gamma = f^*(K_X + \Delta) + E$ as in (3.21.1), then (Y, Γ) is snc and the irreducible components of $\lfloor \Gamma \rfloor$ are disjoint. In particular (Y, Γ) is terminal. Similarly if (X, Δ) is plt, $S = \lfloor \Delta \rfloor$ is irreducible and $T = f_*^{-1} S$, then (Y, Γ) is snc and $(T, (\Gamma - T)|_T)$ is terminal.

3.H Vanishing theorems

Vanishing theorems have played a central role in algebraic geometry for the last couple of decades, especially in classification theory. Kollár [Kol87c] gives an introduction to the basic use of vanishing theorems as well as a survey of results and applications available at the time. For more recent results one should consult [EV92], [Ein97], [Kol97b], [Smi97], [Kov00c], [Kov02], [Kov03a], [Kov03b]. Because of the availability of these surveys, we will only recall statements that are important in this book. However, we should note that the single most crucial ingredient of the vanishing theorems discussed here come from the fact that the Hodge-to-de Rham spectral sequence degenerates at the E_1 level (3.55.4) and its straightforward consequence that the natural map

$$H^i(X, \mathbb{C}) \to H^i(X, \mathscr{O}_X)$$

is surjective (3.56.1) cf. [Kol95, §9].

Any discussion of vanishing theorems should start with the fundamental vanishing theorem of Kodaira.

Theorem 3.37 [Kod53]. *Let X be a smooth complex projective variety and $\mathscr{L}$ an ample line bundle on X. Then*

$$H^i(X, \omega_X \otimes \mathscr{L}) = 0 \qquad \text{for } i > 0.$$

This has been generalized in several ways, but as noted above we will restrict to a select few. The original statement of Kodaira was generalized by Grauert and Riemenschneider to allow semi-ample and big line bundles in place of ample ones.

Theorem 3.38 [GR70]. *Let X be a smooth complex projective variety and $\mathscr{L}$ a semi-ample and big line bundle on X. Then*

$$H^i(X, \omega_X \otimes \mathscr{L}) = 0 \qquad \text{for } i > 0.$$

In fact, probably the most frequently used form of this vanishing theorem is the relative version:

Theorem 3.39 [GR70]. *Let X be a smooth variety, $\phi : X \to Y$ a projective birational morphism, and $\mathscr{L}$ a semi-ample line bundle on X. Then*

$$\mathscr{R}^i\phi_*(\omega_Y \otimes \mathscr{L}) = 0 \qquad \text{for } i > 0.$$

Note that as ϕ is birational, then $\mathscr{L}$ is automatically ϕ-big.

The "semi-ample" hypothesis was later replaced by the weaker "nef" hypothesis by Kawamata and Viehweg.

Theorem 3.40 [Kaw82a, Vie82]. *Let X be a smooth complex projective variety and $\mathscr{L}$ a nef and big line bundle on X. Then*

$$H^i(X, \omega_X \otimes \mathscr{L}) = 0 \qquad \text{for } i > 0.$$

REMARK 3.41. We will need a more general version of this theorem which is stated in (3.45).

Akizuki and Nakano extended Kodaira's vanishing theorem to include other exterior powers of the sheaf of differential forms.

Theorem 3.42 (Akizuki-Nakano [AN54]). *Let X be a smooth complex projective variety and $\mathscr{L}$ an ample line bundle on X. Then*

$$H^q(X, \Omega_X^p \otimes \mathscr{L}) = 0 \qquad \text{for } p + q > \dim X.$$

REMARK 3.43. Ramanujam [Ram72] gave a simplified proof of (3.42) and showed that it does not hold if one only requires $\mathscr{L}$ to be semi-ample and big.

The following generalization of (3.42) to sheaves of logarithmic differential forms, due to Esnault and Viehweg, is also important in many applications.

Theorem 3.44 [EV90, 6.4]. *Let X be a smooth complex projective variety, $\mathscr{L}$ an ample line bundle and D a normal crossing divisor on X. Then*

$$H^q\left(X, \Omega_X^p(\log D) \otimes \mathscr{L}\right) = 0 \qquad \text{for } p + q > \dim X.$$

Extending the known vanishing theorems in a different direction, Navarro-Aznar *et al.* proved a version of the Kodaira-Akizuki-Nakano vanishing theorem for singular varieties that implies the above stated theorems: (3.37), (3.38), and (3.42). For details see [Nav88] in [GNPP88] and Theorem 16.62 in Chapter 16.

Next we recall the general version of the Kawamata-Viehweg vanishing theorem that we will need in the sequel. This is a technical result that is used extensively throughout the minimal model program. Note that it is known to fail in characteristic $p > 0$ and this failure is the main reason why the results described in this book do not readily extend to characteristic $p > 0$.

Theorem 3.45 (Kawamata-Viehweg vanishing). *Let $f: Y \to X$ be a proper morphism of quasi-projective varieties with Y smooth, L a Cartier divisor on Y which is numerically equivalent to $M + \Delta$ where M is an f-nef and f-big $\mathbb{R}$-Cartier divisor and (Y, Δ) is klt. Then*

$$\mathscr{R}^j f_* \mathscr{O}_Y(K_Y + L) = 0 \qquad \text{for } j > 0.$$

This vanishing theorem has many important consequences. Here we recall a few.

Corollary 3.46. *Let $f: Y \to X$ be a projective morphism of quasi-projective varieties with Y smooth, (Y, Γ) a dlt pair and L a Cartier divisor such that $L - (K_Y + \Gamma)$ is f-ample; then*

$$\mathcal{R}^j f_* \mathcal{O}_Y(L) = 0 \qquad \text{for } j > 0.$$

Proof. See [KM98, 2.68]. □

Not suprisingly Kawamata-Viehweg vanishing has interesting applications to the study of the singularities of klt pairs. Perhaps the most well-known result in this direction is a theorem of Elkik (see (15.9)) stating that klt singularities are rational.

Next we recall another consequence of Kawamata-Viehweg vanishing which is used to prove a version of inversion of adjunction.

Theorem 3.47 (Kollár-Shokurov Connectedness). *Let $f : Y \to X$ be a proper birational morphism where Y is smooth and X is normal. If $D = \sum d_i D_i$ is an $\mathbb{R}$-divisor with simple normal crossings support such that $f_* D \geq 0$ and $-(K_Y + D)$ is f-nef, then* $\operatorname{Supp}(\sum_{i:d_i \geq 1} D_i)$ *is connected in a neighborhood of any fiber of f.*

Proof. See [KM98, 5.48]. □

REMARK 3.48. Kollár and Kovács (cf. [KK10, 5.1]) have generalized (3.47) to the context where $f : Y \to X$ is a proper (not necessarily birational) morphism with connected fibers between normal varieties. In this situation, (Y, Δ) is dlt and $K_Y + \Delta \sim_{\mathbb{Q},f} 0$. For an arbitrary $x \in X$ let U denote an étale local neighbourhood of $x \in X$. Then $f^{-1}(U) \cap \operatorname{nklt}(Y, \Delta)$ is either

(3.48.1) connected, or

(3.48.2) has two connected components, both of which dominate U and (Y, Δ) is plt near $f^{-1}(x)$.

The following corollary of (3.47) is known as inversion of adjunction.

Corollary 3.49. *Let $(X, S + B)$ be a pair such that $S \subset X$ is a normal reduced Weil divisor. We write $(K_X + S + B) = K_S + B_S$ as in* (3.24)*. Then $(X, S + B)$ is plt on a neighborhood of S if and only if (S, B_S) is klt.*

Proof. Exercise (or see [KM98, 5.50]). □

REMARK 3.50. By a result of Kawakita (cf. [Kaw07]) a similar result holds for lc pairs: If $(X, S + B)$ is a pair and $S \subset X$ is a reduced Weil divisor with normalization $S^\nu \to S$, then $(X, S + B)$ is lc on a neighborhood of S if and only if (S^ν, B_{S^ν}) is lc where $(K_X + S + B)|_{S^\nu} = K_{S^\nu} + B_{S^\nu}$ as in (3.24).

EXERCISE 3.51. Show that (3.45) implies (3.46).

3.I Rational and Du Bois singularities

These are among the most important classes of singularities, especially for the goals of this manuscript. The essence of rational singularities is that their cohomological behavior is very similar to that of smooth points. For instance, vanishing theorems can be easily extended to varieties with rational singularities. Establishing that a certain class of singularities is rational opens the door to using very powerful tools on varieties with those singularities. Du Bois singularities are probably harder to appreciate at first, but they are equally important. Their main importance comes from two facts: They are not too far from rational singularities, that is, they share many of their properties, but the class of Du Bois singularities is more inclusive than that of rational singularities. For instance, log canonical singularities are Du Bois, but not necessarily rational.

DEFINITION 3.52. Let X be a normal variety and $\phi : Y \to X$ a resolution of singularities. X is said to have **rational** singularities if $\mathcal{R}^i\phi_*\mathcal{O}_Y = 0$ for all $i > 0$, or equivalently if the natural map $\mathcal{O}_X \to \mathcal{R}\phi_*\mathcal{O}_Y$ is a quasi-isomorphism.

A very useful property of rational singularities is that they are Cohen-Macaulay. We will define this notion next.

DEFINITION 3.53. A finitely generated non-zero module M over a noetherian local ring R is called **Cohen-Macaulay** if its depth over R is equal to its dimension. For the definition of depth and dimension we refer the reader to [BH93]. The ring R is called **Cohen-Macaulay** if it is a Cohen-Macaulay module over itself.

Let X be a scheme and $x \in X$ a point. We say that X has **Cohen-Macaulay** singularities at x (or simply X is **CM** at x), if the local ring $\mathcal{O}_{X,x}$ is Cohen-Macaulay.

If in addition, X admits a dualizing sheaf ω_X which is a line bundle in a neighbourhood of x, then X is **Gorenstein** at x.

The scheme X is **Cohen-Macaulay** (resp. **Gorenstein**) if it is **Cohen-Macaulay** (resp. **Gorenstein**) at x for all $x \in X$.

Du Bois singularities are defined via Deligne's Hodge theory and so their strong connection to the singularities discussed above might seem unexpected. Nevertheless, they play a very important role. We will also need a little preparation before we can define these singularities.

The starting point is Du Bois's construction, following Deligne's ideas, of the generalized de Rham complex, which we call the Deligne-Du Bois complex. Recall, that if X is a smooth complex algebraic variety of dimension n, then the sheaves of differential p-forms with the usual exterior differentiation give a resolution of the constant sheaf $\mathbb{C}_X$. I.e., one has a complex of sheaves,

$$\mathcal{O}_X \xrightarrow{d} \Omega_X^1 \xrightarrow{d} \Omega_X^2 \xrightarrow{d} \Omega_X^3 \xrightarrow{d} \cdots \xrightarrow{d} \Omega_X^n \simeq \omega_X,$$

which is quasi-isomorphic to the constant sheaf $\mathbb{C}_X$ via the natural map $\mathbb{C}_X \to \mathcal{O}_X$ given by considering constants as holomorphic functions on X. Recall that this complex **is not** a complex of quasi-coherent sheaves. The sheaves in the complex are quasi-coherent, but the maps between them are not $\mathcal{O}_X$-module morphisms. Notice however that this is actually not a shortcoming; as $\mathbb{C}_X$ is not a quasi-coherent sheaf, one cannot expect a resolution of it in the category of quasi-coherent sheaves.

The Deligne-Du Bois complex is a generalization of the de Rham complex to singular varieties. It is a complex of sheaves on X that is quasi-isomorphic to the constant sheaf, $\mathbb{C}_X$. The terms of this complex are harder to describe but its properties, especially cohomological properties, are very similar to the de Rham complex of smooth varieties. In fact, for a smooth variety the Deligne-Du Bois complex is quasi-isomorphic to the de Rham complex, so it is indeed a direct generalization.

The construction of this complex, $\underline{\Omega}^{\bullet}_X$, is based on simplicial resolutions. The reader interested in the details is referred to the original article [DB81]. Note also that a simplified construction was later obtained in [Car85] and [GNPP88] via the general theory of polyhedral and cubic resolutions. An easily accessible introduction can be found in [Ste85]. Other useful references are the recent book [PS08] and the survey [KS09]. We will actually not use these resolutions here. They are needed for the construction, but if one is willing to believe the listed properties (which follow in a rather straightforward way from the construction), then one should be able follow the material presented here.

Recently Schwede found a simpler alternative construction of (part of) the Deligne-Du Bois complex that does not need a simplicial resolution (3.60). This allows one to define Du Bois singularities (3.54) without needing simplicial resolutions and it is quite useful in applications. For applications of the Deligne-Du Bois complex and Du Bois singularities other than the ones listed here see [Ste83], [Kol95, Chapter 12], [Kov99, Kov00c].

The word "hyperresolution" will refer to either a simplicial, polyhedral, or cubic resolution. Formally, the construction of $\underline{\Omega}^{\bullet}_X$ is the same regardless of the type of resolution used and no specific aspects of either types will be used.

The following definition is included to make sense of the statements of some of the forthcoming theorems. It can be safely ignored if the reader is not interested in the detailed properties of the Deligne-Du Bois complex and is willing to accept that it is a very close analog of the de Rham complex of smooth varieties.

DEFINITION 3.54. Let X be a complex scheme and D a closed subscheme whose complement in X is dense. Then $(X_\bullet, D_\bullet) \to (X, D)$ is a **good hyperresolution** if $X_\bullet \to X$ is a hyperresolution, and if $U_\bullet = X_\bullet \times_X (X \setminus D)$ and $D_\bullet = X_\bullet \setminus U_\bullet$, then D_i is a divisor with normal crossings on X_i for all i. (For a primer on hyperresolutions see the appendix of [KS09].)

Let X be a complex scheme (i.e., a scheme of finite type over $\mathbb{C}$) of dimension n. Let $D_{\text{filt}}(X)$ denote the derived category of filtered complexes of $\mathscr{O}_X$-modules with differentials of order ≤ 1 and $D_{\text{filt,coh}}(X)$ the subcategory of $D_{\text{filt}}(X)$ of complexes $K^\bullet$, such that for all i, the cohomology sheaves of $Gr^i_{\text{filt}} K^\bullet$ are coherent cf. [DB81], [GNPP88]. Let $D(X)$ and $D_{\text{coh}}(X)$ denote the derived categories with the same definition except that the complexes are assumed to have the trivial filtration. The superscripts $+, -, b$ carry the usual meaning (bounded below, bounded above, bounded). Isomorphism in these categories is denoted by $\simeq_{\text{qis}}$. A sheaf $\mathscr{F}$ is also considered as a complex $\mathscr{F}^\bullet$ with $\mathscr{F}^0 = \mathscr{F}$ and $\mathscr{F}^i = 0$ for $i \neq 0$. If $K^\bullet$ is a complex in any of the above categories, then $h^i(K^\bullet)$ denotes the i-th cohomology sheaf of $K^\bullet$.

The right derived functor of an additive functor F, if it exists, is denoted by $\mathscr{R}F$ and $\mathscr{R}^iF$ is short for $h^i \circ \mathscr{R}F$. Furthermore, $\mathbb{H}^i$, $\mathbb{H}^i_Z$, and $\mathscr{H}^i_Z$ will denote $\mathscr{R}^i\Gamma$, $\mathscr{R}^i\Gamma_Z$, and $\mathscr{R}^i\mathscr{H}_Z$ respectively, where Γ is the functor of global sections, Γ_Z is the functor of global sections with support in the closed subset Z, and $\mathscr{H}_Z$ is the functor of the sheaf of local sections with support in the closed subset Z. Note that according to this terminology, if $\phi: Y \to X$ is a morphism and $\mathscr{F}$ is a coherent sheaf on Y, then $\mathscr{R}\phi_*\mathscr{F}$ is the complex whose cohomology sheaves give rise to the usual higher direct images of $\mathscr{F}$.

Theorem 3.55 [DB81, 6.3, 6.5]. *Let X be a complex scheme of finite type and D a closed subscheme whose complement is dense in X. Then there exists a unique object $\underline{\Omega}^\bullet_X(\log D) \in \operatorname{Ob} D_{\mathrm{filt}}(X)$ such that using the notation*

$$\underline{\Omega}^p_X(\log D) := Gr^p_{\mathrm{filt}}\,\underline{\Omega}^\bullet_X(\log D)[p],$$

it satisfies the following properties:

(3.55.1) Let $j: X \setminus D \to X$ be the inclusion map. Then

$$\underline{\Omega}^\bullet_X(\log D) \simeq_{\mathrm{qis}} Rj_*\mathbb{C}_{X\setminus D}.$$

(3.55.2) $\underline{\Omega}^\bullet_{(\_)}(\log(\_))$ is functorial, i.e., if $\phi: Y \to X$ is a morphism of complex schemes of finite type, then there exists a natural map ϕ^* of filtered complexes

$$\phi^*: \underline{\Omega}^\bullet_X(\log D) \to \mathscr{R}\phi_*\underline{\Omega}^\bullet_Y(\log \phi^* D).$$

Furthermore, $\underline{\Omega}^\bullet_X(\log D) \in \operatorname{Ob}\left(D^b_{\mathrm{filt,coh}}(X)\right)$ and if ϕ is proper, then ϕ^* is a morphism in $D^b_{\mathrm{filt,coh}}(X)$.

(3.55.3) Let $U \subseteq X$ be an open subscheme of X. Then

$$\underline{\Omega}^\bullet_X(\log D)\big|_U \simeq_{\mathrm{qis}} \underline{\Omega}^\bullet_U(\log D|_U).$$

(3.55.4) If X is proper, then there exists a spectral sequence degenerating at E_1 and abutting to the singular cohomology of $X \setminus D$:

$$E_1^{pq} = \mathbb{H}^q\left(X, \underline{\Omega}^p_X(\log D)\right) \Rightarrow H^{p+q}(X \setminus D, \mathbb{C}).$$

(3.55.5) If $\varepsilon_\bullet: (X_\bullet, D_\bullet) \to (X, D)$ is a good hyperresolution, then

$$\underline{\Omega}^\bullet_X(\log D) \simeq_{\mathrm{qis}} R\varepsilon_{\bullet *}\Omega^\bullet_{X_\bullet}(\log D_\bullet).$$

In particular, $h^i\left(\underline{\Omega}^p_X(\log D)\right) = 0$ for $i < 0$.

(3.55.6) There exists a natural map, $\mathscr{O}_X \to \underline{\Omega}^0_X(\log D)$, compatible with (3.55.2).

(3.55.7) If X is smooth and D is a normal crossing divisor, then

$$\underline{\Omega}^\bullet_X(\log D) \simeq_{\mathrm{qis}} \Omega^\bullet_X(\log D).$$

In particular,

$$\underline{\Omega}^p_X(\log D) \simeq_{\mathrm{qis}} \Omega^p_X(\log D).$$

(3.55.8) If $\phi: Y \to X$ is a resolution of singularities, then

$$\underline{\Omega}^{\dim X}_X(\log D) \simeq_{\mathrm{qis}} \mathscr{R}\phi_*\omega_Y(\phi^* D).$$

REMARK 3.56. Naturally, one may choose $D = \emptyset$ and then it is simply omitted from the notation. The same applies to $\underline{\Omega}^p_X := Gr^p_{\text{filt}} \underline{\Omega}^\bullet_X[p]$.

It turns out that the Deligne-Du Bois complex behaves very much like the de Rham complex for smooth varieties. Observe that (3.55.4) says that the Hodge-to-de Rham spectral sequence works for singular varieties if one uses the Deligne-Du Bois complex in place of the de Rham complex. This has far-reaching consequences and if the associated graded pieces $\underline{\Omega}^p_X(\log D)$ turn out to be computable, then this single property leads to many applications.

Notice that in the case when $D = \emptyset$, (3.55.6) gives a natural map $\mathscr{O}_X \to \underline{\Omega}^0_X$, and we will be interested in situations when this map is a quasi-isomorphism. When X is proper over $\mathbb{C}$, such quasi-isomorphism will imply that the natural map

$$H^i(X^{\text{an}}, \mathbb{C}) \to H^i(X, \mathscr{O}_X) = \mathbb{H}^i(X, \underline{\Omega}^0_X) \tag{3.56.1}$$

is surjective because of the degeneration at E_1 of the spectral sequence in (3.55.4). Notice that this is the condition that is crucial for Kodaira-type vanishing theorems cf. (3.H), [Kol95, §9].

Following Du Bois, Steenbrink was the first to study this condition and he christened this property after Du Bois. It should be noted that many of the ideas that play important roles in this theory originated from Deligne. Unfortunately the now standard terminology does not reflect this.

DEFINITION 3.57. A scheme X is said to have **Du Bois** singularities (or **DB** singularities for short) if the natural map $\mathscr{O}_X \to \underline{\Omega}^0_X$ from (3.55.6) is a quasi-isomorphism.

REMARK 3.58. If $\varepsilon : X_\bullet \to X$ is a hyperresolution of X, then X has Du Bois singularities if and only if the natural map $\mathscr{O}_X \to \mathcal{R}\varepsilon_{\bullet *}\mathscr{O}_{X_\bullet}$ is a quasi-isomorphism.

EXAMPLE 3.59. It is easy to see that smooth points are Du Bois and Deligne proved that normal crossing singularities are Du Bois as well cf. [DJ74, Lemme 2(b)].

We will see more examples of Du Bois singularities in later sections, especially in Chapter 15.

We finish this section with Schwede's characterization of DB singularities. This condition makes it possible to define DB singularities without hyperresolutions, derived categories, etc. This makes it easier to get acquainted with these singularities, but it is still useful to know the original definition for many applications.

Theorem 3.60 [Sch07]. *Let X be a reduced separated scheme of finite type over a field of characteristic zero. Assume that $X \subseteq Y$ where Y is smooth and let $\pi : \widetilde{Y} \to Y$ be a proper birational map with $\widetilde{Y}$ smooth and where $\overline{X} = \pi^{-1}(X)_{\text{red}}$, the reduced pre-image of X, is a simple normal crossings divisor (or in fact any scheme with DB singularities). Then X has DB singularities if and only if the natural map $\mathscr{O}_X \to \mathcal{R}\pi_*\mathscr{O}_{\overline{X}}$ is a quasi-isomorphism.*

In fact, we can say more. There is a quasi-isomorphism $\mathcal{R}\pi_*\mathscr{O}_{\overline{X}} \xrightarrow{\simeq_{\text{qis}}} \underline{\Omega}^0_X$ such that the natural map $\mathscr{O}_X \to \underline{\Omega}^0_X$ can be identified with the natural map $\mathscr{O}_X \to \mathcal{R}\pi_*\mathscr{O}_{\overline{X}}$.

Part II

Recent advances in the minimal model program

CHAPTER 4

Introduction

Let X be a smooth complex projective variety. The minimal model program aims to show that if K_X is pseudo-effective (respectively if K_X is not pseudo-effective), then there exists a finite sequence

$$X = X_1 \dashrightarrow X_2 \dashrightarrow \cdots \dashrightarrow X_n = \bar{X}$$

of well-understood birational maps known as flips and divisorial contractions such that $\bar{X}$ is a minimal model, i.e., $K_{\bar{X}}$ is nef (respectively $\bar{X}$ has the structure of a Mori fiber space, i.e., there is a morphism $f : \bar{X} \to Z$ such that $-K_{\bar{X}}$ is relatively ample over Z).

The main features of the minimal model program were understood in the 1980's by work of S. Mori, Y. Kawamata, J. Kollár, M. Reid, V. Shokurov and others. In order to produce the above sequence of birational maps $X_i \dashrightarrow X_{i+1}$, one proceeds as follows:

- If K_{X_i} is nef, then X_i is a minimal model and there is nothing to do.
- If K_{X_i} is not nef, then by the Cone theorem, there exists a negative extremal ray and a corresponding morphism $f_i : X_i \to Z_i$ (of normal varieties, surjective with connected fibers and $\varrho(X_i/Z_i) = 1$) such that $-K_{X_i}$ is ample over Z_i.
- If $\dim Z_i < \dim X_i$, then $X_i \to Z_i$ is a Mori fiber space and we are done.
- If $\dim Z_i = \dim X_i$ and $\dim \operatorname{Ex}(f_i) = \dim X_i - 1$, then this is a divisorial contraction and we let $X_{i+1} = Z_i$.
- If $\dim Z_i = \dim X_i$ and $\dim \operatorname{Ex}(f_i) < \dim X_i - 1$, then this is a flipping contraction. In this case, Z_i has bad singularities and we may not let $X_{i+1} = Z_i$. Instead we let X_{i+1} be the flip of f_i cf. (5.14) (assuming that this exists cf. (5.73)).

In order to complete the minimal model program, it is therefore necessary to prove that flips exist and to show that there is no infinite sequence of flips. In dimension 3, the existence of flips was proved by S. Mori [Mor88], and the termination of flips is straightforward. In higher dimensions, the existence and termination of flips have proven to be difficult problems. In [Sho03], V. Shokurov proved the existence of flips in dimension 4. Following the ideas of V. Shokurov and Y.-T. Siu, C. Hacon and J. McKernan (cf. [HM07]) showed that, assuming the minimal model program in dimension $n-1$ (for varieties of log-general type), then pl-flips exist in dimension n cf. (5.64). Finally C. Birkar, P. Cascini, C. Hacon and J. McKernan (cf. [BCHM10]) showed that, assuming that pl-flips exist in dimension n, then the minimal model program for varieties of log-general type holds in dimension n. In particular it follows that flips exist in all dimensions and that if X is a smooth complex projective variety, then its canonical ring $R(K_X) = \bigoplus_{m \geq 0} H^0(\mathcal{O}_X(mK_X))$ is finitely generated (this was also established for varieties of general type, using analytic methods, by Y.-T. Siu [YT06]). We remark that in [BCHM10] it is shown that sequences of flips for the minimal model program with scaling terminate. This is sufficient to show that minimal models exist, however the problem of termination of an arbitrary sequence of flips remains open.

The main purpose of this part of the book is to give a proof of the results of [HM07] and [BCHM10]. In particular we will show that flips exist (in all dimensions for any klt pair (X, Δ)) and that minimal models exist for any klt pair (X, Δ) such that $K_X + \Delta$ is big. In these lectures we will follow the approach of [HM08] and [BCHM10].

REMARK 4.1. We make no attempt to give a historical account of the various developments that have led to the results mentioned above. We simply emphasize once again that the results of [HM07] and [BCHM10] were deeply influenced by the work of V. Shokurov, Y.-T. Siu and others and heavily rely on the techniques of the minimal model program developed by S. Mori, Y. Kawamata, J. Kollár, M. Reid, V. Shokurov and others. For the convenience of the reader we list several papers and books on the minimal model program: [Kol91], [Deb01], [Mat02], [KM98], [CKM88], [McK07], [KMM87], [Kol92], [Kaw08], [CHK$^+$07], [CKL08].

CHAPTER 5

The main result

5.A The Cone and Basepoint-free theorems

We recall the Cone and Basepoint-free theorems. We refer the reader to [KM98] for a particularly clear exposition.

Theorem 5.1 (Basepoint-free theorem). *Let $f: X \to Z$ be a proper morphism and (X, Δ) be a klt pair. If D is an f-nef Cartier divisor such that $aD-(K_X+\Delta)$ is f-nef and f-big for some $a > 0$, then $|bD|$ is f-free for all $b \gg 0$ (i.e., $f^* f_* \mathcal{O}_X(bD) \to \mathcal{O}_X(bD)$ is surjective).*

REMARK 5.2. Note that if bD is free for all $b \gg 0$, then $D = (b+1)D - bD$ is Cartier and D is nef, so these two conditions are necessary. On the other hand, they are not sufficient as shown by the example of a non-torsion topologically trivial line bundle $P \in \mathrm{Pic}^0(A)$ on an abelian variety A.

Theorem 5.3 (Kollár's effective basepoint-free theorem). *Let $f : X \to Z$ be a projective morphism of normal quasi-projective varieties, (X, Δ) be a klt pair of dimension n, D be an f-nef Cartier divisor such that $aD - (K_X + \Delta)$ is f-nef and f-big for some positive real number $a \geq 0$.*

Then there is an integer $m > 0$ (depending only on a and $\dim X$*) such that mD is f-free.*

Proof. [Kol93]. □

Theorem 5.4 (Cone theorem). *Let $f: X \to Z$ be a projective morphism and (X, Δ) be a klt pair. Then:*

(5.4.1) *There exist countably many rational curves $C_j \subset X$ contracted by f such that $0 < -(K_X+\Delta)\cdot C_j \le 2\dim X$, and*

$$\overline{\mathrm{NE}}(X/Z) = \overline{\mathrm{NE}}(X/Z)_{(K_X+\Delta)\ge 0} + \sum \mathbb{R}_{\ge 0}[C_j].$$

(5.4.2) *For any $\varepsilon > 0$ and any f-ample divisor H,*

$$\overline{\mathrm{NE}}(X/Z) = \overline{\mathrm{NE}}(X/Z)_{(K_X+\Delta+\varepsilon H)\ge 0} + \sum_{\text{finite}} \mathbb{R}_{\ge 0}[C_j].$$

(5.4.3) *If $F \subset \overline{\mathrm{NE}}(X/Z)_{(K_X+\Delta)<0}$ is a $(K_X+\Delta)$-negative extremal face, then there is a unique morphism $\mathrm{cont}_F\colon X \to W$ to a projective variety over Z, such that $(\mathrm{cont}_F)_*\mathcal{O}_X = \mathcal{O}_W$ and such that an irreducible curve $C \subset X$ is contracted by cont_F if and only if $[C] \in F$.*

(5.4.4) *If L is a Cartier Divisor on X such that $L\cdot C = 0$ for any curve C with $[C]\in F$, then $L \sim \mathrm{cont}_F^* L_W$ for some Cartier divisor L_W on W.*

Recall that $N_1(X/Z)= \{\sum a_i C_i : a_i \in \mathbb{R},\ f_*C_i = 0\}/\equiv_Z$ denotes the $\mathbb{R}$-vector space of 1-cycles modulo numerical equivalence over Z, that $\mathrm{NE}(X/Z) = \{\sum a_i[C_i] : a_i \ge 0,\ f * C_i = 0\} \subset N_1(X/Z)$, $\overline{\mathrm{NE}}(X/Z)$ is the closure of $\mathrm{NE}(X/Z)$ in $N_1(X/Z)$, and that if D is an $\mathbb{R}$-Cartier divisor, then $\overline{\mathrm{NE}}(X/Z)_{D\ge 0} = \{x \in \overline{\mathrm{NE}}(X/Z) : D\cdot x \ge 0\}$.

Often one says that $\overline{\mathrm{NE}}(X/Z)_{(K_X+\Delta)<0}$ is locally polyhedral even though this is not actually the case.

EXAMPLE 5.5. Let X be a surface obtained by blowing up the nine intersection points of two general cubic curves on $\mathbb{P}^2$. The pencil of the two cubics becomes basepoint-free on X and defines an elliptic fibration of X over $\mathbb{P}^1$. The nine exceptional curves of the blow-up are sections of this fibration. Choosing one of these sections as the origin, X becomes a group scheme over $\mathbb{P}^1$. Each of the remaining eight sections define automorphisms of $X/\mathbb{P}^1$ by translations (which fix all the fibers). We may assume that there is at least one point among the nine whose order on at least one of the cubics is infinite. Then the orbits of the corresponding automorphism (at least on that fiber) are all infinite. Therefore, there are infinitely many images of the eight exceptional curves on X and we conclude that X contains infinitely many (-1)-curves. It is not hard to see (cf. Castelnuovo's theorem [Har77, 5.7]) that a (-1)-curve on a surface is always a K-negative extremal ray, so we conclude that $\overline{\mathrm{NE}}(X)$ contains infinitely many K-negative extremal rays.

REMARK 5.6. The log canonical version of the Cone theorem is discussed in [Amb03] and [Fuj08a].

EXERCISE 5.7. Deduce the last two statements of (5.4) from (5.1).

EXERCISE 5.8. Show that if (X,Δ) is a klt pair such that Δ is big, then there are finitely many $(K_X+\Delta)$-negative extremal rays.

EXERCISE 5.9. Using the Basepoint-free and Cone theorems for $\mathbb{Q}$-divisors, show that: *If $f : X \to Z$ is a projective morphism and (X,Δ) is a klt pair, D is an f-nef $\mathbb{R}$-Cartier divisor such that $aD - (K_X+\Delta)$ is f-nef and f-big, then D is semiample over Z.*

EXERCISE 5.10. Show that if (X, Δ) is a dlt pair such that $K_X + \Delta$ is nef, $\Delta = A + B$ and $\mathbf{B}_+(A)$ contains no non-klt centers of (X, Δ), then $K_X+\Delta$ is semiample. In particular if (X, Δ) is klt, $K_X + \Delta$ is nef and Δ is big, then $K_X + \Delta$ is semiample.

EXERCISE 5.11. Using the Basepoint-free and Cone theorems for $\mathbb{Q}$-divisors, show that if (X, Δ) is klt and $K_X + \Delta$ is big and nef, then $K_X + \Delta$ is semiample.

EXERCISE 5.12. Let X be the product of two elliptic curves, then $\varrho(X) = 3$ and $N^1(X)$ is generated by f_1, f_2 and δ where f_i denote the two fibers and δ denotes the diagonal. Show that $\overline{\mathrm{NE}}(X)$ is given by the circular cone consisting of curves $D = \sum d_i D_i \in N^1_{\mathbb{R}}(X)$ such that $D^2 \geq 0$ and $D \cdot A \geq 0$ for any ample divisor A.

EXERCISE 5.13. Let $g : X \to \mathbb{P}^2$ be the blow-up of $\mathbb{P}^2$ at 12 points $p_1, \dots, p_{12}$ on a smooth plane cubic $C \subset \mathbb{P}^2$. Show that the strict transform C' of C has self intersection -3 and if the points $p_1, \dots, p_{12}$ are given (resp. are not given) by intersecting C with a quartic, then there is (resp. there can't be) a proper morphism $f : X \to Y$ into a projective space. Note that in both cases, $f : X \to Y$ exists as an analytic morphism.

5.B Flips and divisorial contractions

Definition 5.14. Let $f : X \to Z$ be a projective morphism of normal quasi-projective varieties such that $f_*\mathcal{O}_X = \mathcal{O}_Z$ and $\varrho(X/Z) = 1$. Let (X, Δ) be a $\mathbb{Q}$-factorial dlt pair such that $-(K_X + \Delta)$ is f-ample.

(5.14.1) If $\dim Z < \dim X$, then f is a **Mori fiber space**;

(5.14.2) if $\dim Z = \dim X$ and $\dim \mathrm{Ex}(f) = \dim X - 1$, then f is a **divisorial contraction**;

(5.14.3) if $\dim Z = \dim X$ and $\dim \mathrm{Ex}(f) < \dim X - 1$, then f is a **flipping contraction**.

REMARK 5.15. Note that if (X, Δ) is as above, then there exists a klt pair (X, Δ') such that $\Delta' \in \mathrm{WDiv}_{\mathbb{Q}}(X)$ and $-(K_X + \Delta')$ is f-ample.

REMARK 5.16. Let (X, Δ) be a klt pair and $R = \mathbb{R}_{\geq 0}[\Sigma]$ a $(K_X+\Delta)$-**negative extremal ray**, i.e., a 1-dimensional face of $\overline{\mathrm{NE}}(X)$ such that $(K_X + \Delta) \cdot \Sigma < 0$. By the Cone theorem (5.4), there exists a morphism $f : X \to Z$ that is a Mori fiber space or a flipping contraction or a divisorial contraction.

REMARK 5.17. Let $\pi : X \to U$ be a projective morphism and $f : X \to Z$ be a flipping or divisorial contraction. If the induced rational map $Z \dashrightarrow U$ is a morphism, then we say that f is a flipping or divisorial contraction over U. Note that if f is given by a negative extremal ray $R = \mathbb{R}_{\geq 0}[\Sigma]$, then f is a flipping or divisorial contraction over U if and only if Σ is contracted by π. In this case, Σ spans a negative extremal ray for $\overline{\mathrm{NE}}(X/U)$.

Definition 5.18. Let $f : X \to Z$ be a flipping contraction for the $\mathbb{Q}$-factorial dlt pair (X, Δ). The **flip** $f^+ : X^+ \to Z$ of f is a small birational morphism such that $f^+_*\mathcal{O}_{X^+} = \mathcal{O}_Z$, $\varrho(X^+/Z) = 1$ and $K_{X^+} + \Delta^+$ is f^+-ample where Δ^+ is the strict transform of Δ on X^+.

Recall that a **small morphism** is a birational morphism $f : X \to Z$ such that $\dim \mathrm{Ex}(f) < \dim X - 1$. Note that if f is small and $D \in \mathrm{WDiv}(X)$, then $f_*\mathcal{O}_X(D) = \mathcal{O}_Z(f_*D)$ cf. (5.35).

REMARK 5.19. The **flipping locus** is given by $\mathrm{Ex}(f)$ and the **flipped locus** is $\mathrm{Ex}(f^+)$. It will follow from the proof of (5.21) that $X \dashrightarrow X^+$ induces an isomorphism on the complements of the flipped and flipping loci.

REMARK 5.20. Note that if $f : X \to Z$ is a flipping contraction for a klt pair (X, Δ), then there exists $\Delta' \in \mathrm{WDiv}_{\mathbb{Q}}(X)$ such that (X, Δ') is klt and f is a flipping contraction for (X, Δ'). Therefore, to prove that klt (or dlt) flips exist, it suffices to show that flips exist for klt pairs (X, Δ) with $\Delta \in \mathrm{WDiv}_{\mathbb{Q}}(X)$.

Proposition 5.21. *Let $f : X \to Z$ be a flipping contraction for a dlt pair (X, Δ) such that $\Delta \in \mathrm{WDiv}_{\mathbb{Q}}(X)$. The flip f^+ exists, if and only if $\bigoplus_{m\geq 0} f_*\mathcal{O}_X(m(K_X + \Delta))$ is a finitely generated sheaf of $\mathcal{O}_Z$-algebras. We then have that*

$$X^+ = \mathrm{Proj}_Z \bigoplus_{m\geq 0} f_*\mathcal{O}_X(m(K_X + \Delta)).$$

Proof. Suppose that the flip f^+ exists. Since $X \dashrightarrow X^+$ is small, by (5.35), we have that

$$\bigoplus_{m\geq 0} f_*\mathcal{O}_X(m(K_X + \Delta)) \simeq \bigoplus_{m\geq 0} f^+_*\mathcal{O}_{X^+}(m(K_{X^+} + \Delta^+)).$$

Since $K_{X^+} + \Delta^+$ is ample over Z, the right-hand side is a finitely generated sheaf of $\mathcal{O}_Z$-algebras and $X^+ = \mathrm{Proj}_Z \bigoplus_{m\geq 0} f_*\mathcal{O}_X(m(K_X + \Delta))$.

Suppose now that $\bigoplus_{m\geq 0} f_*\mathcal{O}_X(m(K_X + \Delta))$ is a finitely generated $\mathcal{O}_Z$-algebra. Let $X^+ = \mathrm{Proj}_Z \bigoplus_{m\geq 0} f_*\mathcal{O}_X(m(K_X + \Delta))$. Clearly X and X^+ are isomorphic over $Z - \mathrm{Ex}(f)$. We may assume that

$$f^+_*\mathcal{O}_{X^+}(1) \simeq f_*\mathcal{O}_X(m(K_X + \Delta)) \simeq \mathcal{O}_Z(m(K_Z + f_*\Delta))$$

for some integer $m > 0$ (where the second isomorphism follows by (5.35)). Let $E \subset X^+$ be an exceptional divisor for $f^+ : X^+ \to Z$. For any integer $t \gg 0$, we have that

$$\mathcal{O}_Z(tm(K_Z + f_*\Delta)) = f^+_*\mathcal{O}_{X^+}(t) \subsetneq f^+_*\mathcal{O}_{X^+}(t)(E).$$

As $\mathcal{O}_{X^+}(1)$ is ample over Z, the last inclusion is strict for $t \gg 0$, but as $\mathcal{O}_Z(tm(K_Z + f_*\Delta))$ is reflexive and E is exceptional, we have an inclusion $f^+_*\mathcal{O}_{X^+}(t)(E) \hookrightarrow \mathcal{O}_Z(tm(K_Z + f_*\Delta))$. This is a contradiction and hence f^+ is small so that $X \dashrightarrow X^+$ is an isomorphism in codimension 1. It follows that $\varrho(X^+/Z) = \varrho(X/Z) = 1$ (cf. Section 2.B) and that f^+ is the flip of f. □

REMARK 5.22. Note that to prove the existence of the flip f^+, one may work locally over Z. If we assume that $Z = \mathrm{Spec}(A)$ is affine and $\Delta \in \mathrm{WDiv}_{\mathbb{Q}}(X)$, then it suffices to show that

$$R(K_X + \Delta) = \bigoplus_{m\geq 0} H^0(\mathcal{O}_X(m(K_X + \Delta)))$$

is a finitely generated A-algebra. Note that as A is a finitely generated $\mathbb{C}$-algebra, this is equivalent to requiring that $R(K_X + \Delta)$ be a finitely generated $\mathbb{C}$-algebra.

Lemma 5.23 (Negativity lemma). *Let $f: Y \to X$ be a proper birational morphism of normal quasi-projective varieties. If $-B$ is an f-nef $\mathbb{R}$-Cartier divisor on Y, then*

(5.23.1) $B \geq 0$ *if and only if* $f_* B \geq 0$.

(5.23.2) If $B \geq 0$ *and* $x \in X$, *then either* $f^{-1}(x) \subset \operatorname{Supp}(B)$ *or* $f^{-1}(x) \cap \operatorname{Supp}(B) = \emptyset$.

Proof. See [KM98, 3.39]. □

Lemma 5.24. *Let $f: X \to Z$ be a flipping contraction and $f^+: X^+ \to Z$ be its flip. Then X^+ is $\mathbb{Q}$-factorial and $a(E; X, \Delta) \leq a(E; X^+, \Delta^+)$ for any divisor E over X and the strict inequality holds if the center of E is contained in the flipping or flipped locus. In particular, if (X, Δ) is terminal, canonical, lc, klt, plt or dlt, then so is (X^+, Δ^+).*

Proof. Exercise (use (5.23) or see [KM98, 3.42, 3.44]). □

Lemma 5.25. *Let $f: X \to Z$ be a divisorial contraction. Show that Z is $\mathbb{Q}$-factorial and $a(E; X, \Delta) \leq a(E; Z, f_*\Delta)$ for any divisor E over X and that strict inequality holds if the center of E is contained in* $\operatorname{Ex}(f)$. *In particular, if (X, Δ) is lc, klt, plt or dlt, then so is $(Z, f_*\Delta)$. There are examples where (X, Δ) is terminal, but $(Z, f_*\Delta)$ is not terminal. However, if (X, Δ) is terminal resp. canonical and* $\operatorname{Supp}(E) \wedge \operatorname{Supp}(\Delta) = 0$, *then $(Z, f_*\Delta)$ is terminal resp. canonical.*

Proof. Exercise, see [KM98, 3.43, 3.44]. □

Definition 5.26. Let (X, Δ) be a dlt pair and $\pi : X \to U$ a projective morphism of normal quasi-projective varieties. A **minimal model program** or **mmp** for (X, Δ) over U is a sequence of birational maps $\phi_i : X_i \dashrightarrow X_{i+1}$ with $i \in I$ such that $X_0 = X$, Δ_i is the strict transform of Δ, $X_i \dashrightarrow X_{i+1}$ is a flip or divisorial contraction for $K_{X_i} + \Delta_i$ over U corresponding to a $(K_{X_i} + \Delta_i)$-negative extremal ray $R_i = \mathbb{R}_{\geq 0}[\Sigma_i]$ in $\overline{\mathrm{NE}}(X_i/U)$ (in particular $\pi_i : X_i \to U$ is a projective morphism) and either

(5.26.1) I is infinite, i.e., $I = \mathbb{Z}_{\geq 0}$; or

(5.26.2) $I = \{0, 1, \ldots, N-1\}$ and $K_{X_N} + \Delta_N$ is nef over U; or

(5.26.3) $I = \{0, 1, \ldots, N-1\}$ and there is a $(K_{X_N} + \Delta_N)$-Mori fiber space $X_N \to Z$ over U.

REMARK 5.27. Note that if ϕ_i is a flip, then $\varrho(X_i) = \varrho(X_{i+1})$ and if ϕ_i is a divisorial contraction, then $\varrho(X_i) = \varrho(X_{i+1}) + 1$. Since $\varrho(X_i) \in \mathbb{Z}_{>0}$, it follows that we have at most $\varrho(X) - 1$ divisorial contractions. To show that a given mmp terminates, it suffices to show that there is no infinite sequence of flips. To show that we may run the mmp (for any klt pair (X, Δ)) we must show that flips exist (cf. (5.73)).

REMARK 5.28. We will also refer to the (X, Δ)-mmp as the $(K_X + \Delta)$-mmp.

Definition 5.29. Let $\pi : X \to U$ and $\pi' : Y \to U$ be projective morphisms of normal quasi-projective varieties. Let (X, Δ) be a dlt pair and $\phi : X \dashrightarrow Y$ a rational map over U such that

(5.29.1) ϕ^{-1} contracts no divisors;
(5.29.2) Y is $\mathbb{Q}$-factorial;
(5.29.3) $K_Y + \phi_*\Delta$ is nef over U; and
(5.29.4) $a(F; X, \Delta) < a(F; Y, \phi_*\Delta)$ for any ϕ-exceptional prime divisor $F \subset X$.
Then ϕ is a **log terminal model (ltm** for short) for (X, Δ) over U. If $K_X + \Delta$ is nef over U, we will say that (X, Δ) is a **log terminal model**.

REMARK 5.30. Assume that $U = \operatorname{Spec}(\mathbb{C})$. Notice that by the Negativity lemma (5.23), $a(F; X, \Delta) \leq a(F; Y, \phi_*\Delta)$ for any divisor F over Y. In fact, if $p : W \to X$ and $q : W \to Y$ is a common resolution, then

$$p^*(K_X + \Delta) = q^*(K_Y + \phi_*\Delta) + E$$

where

$$E = \sum (a(F; Y, \phi_*\Delta) - a(F; X, \Delta))F$$

is an effective and q-exceptional $\mathbb{R}$-divisor. Moreover, if $k(K_X + \Delta)$ and $k(K_Y + \phi_*\Delta)$ are Cartier, then

$$H^0(\mathcal{O}_X(k(K_X + \Delta))) \simeq H^0(\mathcal{O}_Y(k(K_Y + \phi_*\Delta))).$$

Notice that if $K_Y + \phi_*\Delta$ is semiample, then $\mathbf{Fix}(K_X + \Delta) = \operatorname{Supp}(p_*E)$ is the set of all ϕ-exceptional prime divisors $F \subset X$.

Lemma 5.31. *Let $\pi : X \to U$ be a projective morphism of normal quasi-projective varieties and (X, Δ) be a dlt pair. If $\phi_i : X_i \dashrightarrow X_{i+1}$ is a sequence of $(K_X + \Delta)$-flips and divisorial contractions over U for $i \in \{0, 1, \dots, N-1\}$ such that $X = X_0$, $\Delta_{i+1} = (\phi_i)_*\Delta_i$ and $K_{X_N} + \Delta_N$ is nef over U, then the induced rational map $X \dashrightarrow X_N$ is an ltm for (X, Δ) over U.*

Proof. Exercise. □

REMARK 5.32. For a discussion of flips for non-$\mathbb{Q}$-factorial pairs we suggest [Fuj07]. For a discussion of flips for lc pairs, we suggest [Fuj08a].

EXERCISE 5.33. Show that if $f : X \to Z$ is a flipping contraction for a pair (X, Δ), then $K_Z + f_*\Delta$ is not $\mathbb{Q}$-Cartier.

EXERCISE 5.34. Show that if $f : X \to Z$ is a divisorial contraction, then $\operatorname{Ex}(f)$ is irreducible. (This is not the case for flipping contractions! See [Fuj07].)

EXERCISE 5.35. Let $f : X \dashrightarrow Y$ be a small birational map of normal varieties and $D \in \operatorname{WDiv}(X)$. Show that $H^0(\mathcal{O}_X(D)) \simeq H^0(\mathcal{O}_Y(f_*D))$.

EXERCISE 5.36. Let (X, Δ) be a klt pair and $\pi : X \to Z$ be a projective morphism such that Δ is big over Z. Show that the divisors contracted by an ltm of (X, Δ) over Z are the divisors contained in $\mathbf{B}(K_X + \Delta/Z)$.

EXERCISE 5.37. Let (X, Δ) be a klt pair and $\phi_i : X \dashrightarrow Y_i$ be two log terminal models of (X, Δ). Show that Y_1 is isomorphic to Y_2 in codimension 1 and that $a(E; Y_1, \phi_{1*}\Delta) = a(E; Y_2, \phi_{2*}\Delta)$ for any divisor E over Y_i.

EXERCISE 5.38. Let V be an abelian threefold. We regard V as an additive group and we let -1_V be the involution of V given by sending $v \in V$ to $-v$. Let W be the quotient of V by the involution -1_V. Show that $(W, 0)$ is an ltm and that it has canonical singularities (in fact W has 2^6 singular points of index 2 corresponding to the 2^6 fixed points of -1_V). Show that any other ltm of $(W, 0)$ is singular.

EXERCISE 5.39. Let (X, Δ) be a klt pair and $X_i \dashrightarrow X_{i+1}$ a sequence of flips and divisorial contractions for the $(K_X+\Delta)$-mmp. If (X, Δ') is a klt pair such that $K_X+\Delta \equiv \lambda(K_X + \Delta')$ for some $\lambda > 0$, then show that $X_i \dashrightarrow X_{i+1}$ is a sequence of flips and divisorial contractions for the $(K_X + \Delta')$-mmp.

EXERCISE 5.40. Let $\pi : X \to U$ be a projective morphism and (X, Δ) and (X, Δ') be klt pairs such that $K_X + \Delta \equiv_U \lambda(K_X + \Delta')$ for some $\lambda > 0$. Show that if $\phi : X \dashrightarrow Y$ is an ltm for $K_X + \Delta$ over U, then it is an ltm for $K_X + \Delta'$ over U.

EXERCISE 5.41. Assume (5.54). Let (X, Δ) be a klt pair. Let $\mathcal{E} = \{E_i\}$ be a finite collection of exceptional divisors over X such that $a(E_i; X, \Delta) \leq 0$. Show that there is a birational morphism $f : Y \to X$ from a $\mathbb{Q}$-factorial variety such that $K_Y + \Gamma = f^*(K_X + \Delta)$ is klt, nef over X and the set of f-exceptional divisors $F \subset Y$ is $\mathcal{E}$. When $\mathcal{E} = \emptyset$, we say that $Y \to X$ is a $\mathbb{Q}$-factorialization of X and when $\mathcal{E}$ consists of all divisors over X with $a(E; X, \Delta) \leq 0$, we say that Y is a terminalization of (X, Δ).

EXERCISE 5.42. Let (X, Δ) be a klt pair and $\pi : X \to U$ be a projective morphism. Let $f : Y \to X$ be a log resolution of (X, Δ) and write $K_Y + \Gamma = f^*(K_X + \Delta) + E$ as in (3.21.1). Let $F \geq 0$ be an $\mathbb{R}$-divisor whose support equals $\mathrm{Ex}(f)$ and $(Y, \Gamma + F)$ is klt. Show that if $\phi : Y \dashrightarrow Z$ is an ltm model over U for $(Y, \Gamma + F)$, then $X \dashrightarrow Z$ is an ltm model over U for (X, Δ).

EXERCISE 5.43. Let (X, Δ) be a dlt pair such that $\{\Delta\} = A + B$ where $\mathbf{B}_+(A)$ contains no non-klt center of (X, Δ). Let $f : X \dashrightarrow X'$ be a flip or a divisorial contraction and $\Delta' = f_*\Delta$, $A' = f_*A$. Show that $\mathbf{B}_+(A')$ contains no non-klt center of (X', Δ').

EXERCISE 5.44. Let X be a threefold. Assuming the existence of flips, show that if $f : X \to Z$ is a K_X-flipping contraction, then X is singular.

EXERCISE 5.45. Let $f : Y \to X$ be the blow-up of a smooth subvariety of codimension ≥ 2 contained in X_{reg}. Show that f is a (K_Y-negative) divisorial contraction.

EXERCISE 5.46. Let X be a smooth variety and $\mathcal{L}$ be a line bundle on X. Let P be the $\mathbb{P}^1$ bundle over X corresponding to $\mathcal{O}_X \oplus \mathcal{L}$. Show that the induced morphism $P \to X$ is a Mori fiber space. If $X = \mathbb{P}^n$ and $\mathcal{L} = \mathcal{O}_{\mathbb{P}^n}(a)$ with $0 < a < n + 1$, and X_0 is the negative section corresponding to the surjection $\mathcal{O}_{\mathbb{P}^n} \oplus \mathcal{L} \to \mathcal{O}_{\mathbb{P}^n}$, then show that there is a divisorial contraction $P \to C$ where C is a cone over $\mathbb{P}^n$.

EXERCISE 5.47. Let $V \subset \mathrm{Spec}\mathbb{C}[w, x, y, z]$ be the cone over $\mathbb{P}^1 \times \mathbb{P}^1$ defined by $\{wz - xy = 0\}$. Show that there is a resolution $f : W \to V$ that replaces the vertex by a divisor E isomorphic to $\mathbb{P}^1 \times \mathbb{P}^1$ and that there exist morphisms $W \to W_1$ and $W \to W_2$ that contract the first and second rulings of E. Show that the induced morphisms $W_i \to V$ are K_{W_i}-trivial. The map $W_1 \dashrightarrow W_2$ is called a flop. Find $\Delta_1 \in \mathrm{WDiv}_{\mathbb{Q}}(W_1)$ such that $W_1 \dashrightarrow W_2$ is a $(K_{W_1} + \Delta_1)$-flip.

EXERCISE 5.48. Let W be the $\mathbb{P}^1$ bundle over $\mathbb{P}^1 \times \mathbb{P}^1$ defined by $\mathcal{O}_{\mathbb{P}^1 \times \mathbb{P}^1} \oplus (\mathcal{O}_{\mathbb{P}^1}(1) \boxtimes \mathcal{O}_{\mathbb{P}^1}(3))$, Z the cone over $\mathbb{P}^1 \times \mathbb{P}^1$ given by contracting the negative section E of $\mathbb{P}^1 \times \mathbb{P}^1$. Show that the 3-folds given by contracting the 2-rulings of $E \simeq \mathbb{P}^1 \times \mathbb{P}^1$ define a flip over Z.

5.C The minimal model program for surfaces

Let (X, Δ) be a klt surface (i.e., $\dim X = 2$), then there are no $(K_X + \Delta)$-flips and so a $(K_X + \Delta)$-mmp is given by a finite sequence of divisorial contractions $\phi_i : X_i \to X_{i+1}$.

If $\Delta = 0$ and $(X, 0)$ is terminal, then X is smooth and so is X_i for $i \in \{0, 1, \dots, N\}$. Note that then ϕ_i contracts a curve E_i to a point $x_{i+1} \in X_{i+1}$. In fact ϕ_i is the blow-up of $x_{i+1} \in X_{i+1}$ so that $E_i \simeq \mathbb{P}^1$ and $K_{X_i} \cdot E_i = -1$.

If, moreover K_X is pseudo-effective, then $\kappa(X) \geq 0$ and X has a unique ltm X_N such that K_{X_N} is nef. It is in fact known that K_{X_N} is semiample. If $\kappa(X) = 0$, then $12K_{X_N} \sim 0$ and X is an abelian, bi-elliptic, K3 or Enriques surface. If $\kappa(X) = 1$, then there is a morphism to a curve

$$f : X_N \to Z = \operatorname{Proj}(R(K_X))$$

such that $K_{X_N} \sim_{\mathbb{Q}} f^*(K_Z + \Delta)$ where (Z, Δ) is klt and $K_Z + \Delta$ is an ample $\mathbb{Q}$-divisor. If $\kappa(X) = 2$, then $|5K_{X_N}|$ is basepoint-free and the corresponding birational morphism

$$X_N \to X_{\text{can}} = \operatorname{Proj}(R(K_X))$$

is a map to a surface with canonical singularities such that $K_{X_{\text{can}}}$ is ample and $K_{X_N} = f^* K_{X_{\text{can}}}$.

If K_X is not pseudo-effective, then there is a Mori-fiber space $f : X_N \to Z$. If $\dim Z = 0$, then $\varrho(X_N) = 1$ and $-K_{X_N}$ is ample. It follows that $X_N \simeq \mathbb{P}^2$. If $\dim Z = 1$, then Z is a smooth curve of genus $h^0(\Omega^1_{X_N}) = h^0(\Omega^1_X)$ and the general fibers of f are rational curves. Note that for any given X the Mori fiber space $X_N \to Z$ is not uniquely determined.

REMARK 5.49. Some references for the theory of surfaces are [Bea96], [Rei97], [Băd01] and [BHPV04]. The mmp for surfaces in characteristic $p > 0$ is worked out in [KK].

EXERCISE 5.50. Show that if $\kappa(X) = 2$ (or more generally if $R(K_X)$ is finitely generated and $\kappa(X) = \dim(X)$), then $X_{\text{can}} = \operatorname{Proj} \oplus_{m \geq 0} \mathcal{O}_X(mK_X)$ has canonical singularities.

EXERCISE 5.51. Show that there exists a surface X and mmps $X \to X_N$ and $X \to X'_N$ such that $X_N = \mathbb{P}^2$ and $X'_N = \mathbb{P}^1 \times \mathbb{P}^1$ (similarly for $X_N = \mathbb{F}_n$ and $X'_N = \mathbb{F}_{n+1}$ where $\mathbb{F}_n$ is the projective bundle over $\mathbb{P}^1$ corresponding to $\mathcal{O}_{\mathbb{P}^1} \oplus \mathcal{O}_{\mathbb{P}^1}(n)$).

EXERCISE 5.52. Let X be a klt surface. Use (5.41) to show that there exists a minimal resolution $f : Y \to X$ (i.e., Y is smooth and any birational morphism $f' : Y' \to X$ from a smooth surface Y' factors through Y).

EXERCISE 5.53. Let (X, Δ) be a klt surface. Show that X is $\mathbb{Q}$-factorial (in particular X is log terminal).

5.D The main theorem and sketch of proof

Theorem 5.54. *Let $\pi : X \to U$ be a projective morphism of normal quasi-projective varieties, (X, Δ) a $\mathbb{Q}$-factorial klt pair such that Δ is big over U.*

Then there exists a finite sequence of flips and divisorial contractions for the $(K_X + \Delta)$-mmp over U

$$X = X_1 \dashrightarrow X_2 \dashrightarrow \cdots \dashrightarrow X_N$$

such that either $K_{X_N} + \Delta_N$ is nef over U or there exists a morphism $X_N \to Z$ which is a $K_{X_N} + \Delta_N$ Mori fiber space over U.

Proof. See (5.63). □

REMARK 5.55. With this hypothesis, $K_{X_N} + \Delta_N$ is nef over U if and only if $K_X + \Delta$ is pseudo-effective over U and $X_N \to Z$ is a $K_{X_N} + \Delta_N$ Mori fiber space over U if and only if $K_X + \Delta$ is not pseudo-effective over U.

The proof will be done by induction on the dimension and will be subdivided into the following five theorems.

Theorem 5.56 (Existence of pl-flips). *Let (X, Δ) be a plt pair and $f : X \to Z$ be a pl-flipping contraction.*

Then the flip $f^+ : X^+ \to Z$ of f exists.

Theorem 5.57 (Existence of log terminal models). *Let $\pi : X \to U$ be a projective morphism of normal quasi-projective varieties. Let (X, Δ) be a klt pair and D be an effective $\mathbb{R}$-divisor such that Δ is big over U and $K_X + \Delta \sim_{\mathbb{R},U} D$.*

Then there exists an ltm for (X, Δ) over U.

Theorem 5.58 (Non-vanishing theorem). *Let $\pi : X \to U$ be a projective morphism of normal quasi-projective varieties. Let (X, Δ) be a klt pair such that Δ is big over U.*

If $K_X + \Delta$ is pseudo-effective over U, then there exists an effective $\mathbb{R}$-divisor D such that $K_X + \Delta \sim_{\mathbb{R},U} D$.

Theorem 5.59 (Finiteness of models). *Let $\pi : X \to U$ be a projective morphism of normal quasi-projective varieties. Let $\mathcal{C} \subset \mathrm{WDiv}_{\mathbb{R}}(X)$ be a rational polytope such that for any $K_X + \Delta \in \mathcal{C}$, Δ is big over U and (X, Δ) is klt.*

Then there exist finitely many $\phi_i : X \dashrightarrow Y_i$, $1 \le i \le k$, birational maps over U such that if $K_X + \Delta \in \mathcal{C}$ and $K_X + \Delta$ is pseudo-effective over U, then:

(5.59.1) There exists an index $1 \le j \le k$ such that ϕ_j is an ltm of (X, Δ) over U.

(5.59.2) If $\phi : X \dashrightarrow Y$ is an ltm of (X, Δ) over U, then there exists an index $1 \le j \le k$ such that the rational map $\phi_j \circ \phi^{-1} : Y \dashrightarrow Y_j$ is an isomorphism.

Theorem 5.60 (Zariski decomposition). *Let $f: X \to Z$ be a projective morphism to a normal affine variety Z. Let (X, Δ) be a klt pair where $K_X + \Delta$ is pseudo-effective, $\Delta = A + B$ and $A \ge 0$ is an ample $\mathbb{Q}$-divisor and $B \ge 0$. Then:*

(5.60.1) (X, Δ) has an ltm $\mu: X \dashrightarrow Y$ over Z. In particular if $K_X + \Delta$ is $\mathbb{Q}$-Cartier, then $R(K_X + \Delta)$ is finitely generated.

(5.60.2) *Let* $V \subset \mathrm{WDiv}_{\mathbb{R}}(X)$ *be a finite dimensional affine subspace of* $\mathrm{WDiv}_{\mathbb{R}}(X)$ *containing* Δ *which is defined over the rationals. There exists a constant* $\delta > 0$ *such that if* P *is a prime divisor contained in* $\mathbf{B}(K_X + \Delta)$*, then* P *is contained in* $\mathbf{B}(K_X + \Delta')$*, for any* $\mathbb{R}$*-divisor* $\Delta' \in V$ *with* $\|\Delta - \Delta'\| \leq \delta$.

(5.60.3) *Let* $W \subset \mathrm{WDiv}_{\mathbb{R}}(X)$ *be the smallest affine subspace containing* Δ *which is defined over the rationals. Then there is a real number* $\eta > 0$ *and an integer* $r > 0$ *such that if* $\Delta' \in W$, $\|\Delta - \Delta'\| \leq \eta$ *and* $k > 0$ *is an integer such that* $k(K_X + \Delta')/r$ *is Cartier, then* $|k(K_X + \Delta')| \neq \emptyset$ *and every component of* $\mathrm{Fix}(k(K_X + \Delta'))$ *is a component of* $\mathbf{B}(K_X + \Delta)$.

Proof of Theorems 5.56, 5.57, 5.58, 5.59, 5.60. The proof is by induction on the dimension $n = \dim X$. The case $n = 1$ is clear. We will show that Theorems 5.56, 5.57, 5.58, 5.59, 5.60 in dimensions $\leq n - 1$ imply Theorem 5.56 in dimension n cf. (5.69), which implies Theorem 5.57 in dimension n cf. (8.1), which implies Theorem 5.58 in dimension n cf. (9.15), which implies Theorem 5.59 in dimension n cf. (10.1) and (10.2), which implies Theorem 5.60 in dimension n cf. (10.3). □

In the next two sections we will explain two of the main techniques involved in the inductive proof of Theorems 5.56, 5.57, 5.58, 5.59 and 5.60. The first technique is known as the minimal model program with scaling. In this version of the mmp, the choice of extremal rays (and hence the sequence of flips and contractions $X = X_0 \dashrightarrow X_1 \dashrightarrow X_2 \dashrightarrow \cdots$) is (almost) uniquely determined by the initial choice of a divisor H such that $K_X + \Delta + H$ is nef. Moreover, each X_i comes with the additional property that $K_{X_i} + \Delta_i + t_i H_i$ is nef for some $0 \leq t_i \leq 1$. In other words X_i is an ltm for $K_X + \Delta + t_i H$. The termination of a sequence of flips for the mmp with scaling will then follow from the fact that (if Δ is big) there are only finitely many ltms for $K_X + \Delta + tH$ where $t \in [0, 1]$. Note that, we are unable to prove termination of an arbitrary sequence of flips. However, see [KMM87], [Fuj05], [AHK07], [Sho04] and [Sho06] for several partial results.

The second technique known as the reduction to pl-flips is due to V. Shokurov. The main idea is that it suffices to show that flips exist for a special class of flipping contractions known as pl-flipping contractions (or *pre-limiting* flipping contractions). These are flipping contractions $f : X \to Z$ for a plt pair (X, Δ) where $\Delta \in \mathrm{WDiv}_{\mathbb{Q}}(X)$, $\lfloor \Delta \rfloor = S$ is irreducible and $-S$ is f-ample. The advantage is that in order to show that $\oplus_{m \geq 0} f_* \mathcal{O}_X(m(K_X + \Delta))$ is a finitely generated sheaf of $\mathcal{O}_Z$-algebras (and hence that the flip exists), it suffices to show that the restricted algebra, i.e., the image of the restriction homomorphism

$$\bigoplus_{m \geq 0} f_* \mathcal{O}_X(m(K_X + \Delta)) \to \bigoplus_{m \geq 0} f_* \mathcal{O}_S(m(K_X + \Delta)|_S)$$

is a finitely generated sheaf of $\mathcal{O}_Z$-algebras. Since $\dim(S) = \dim(X) - 1$ this lends itself to a proof by induction on the dimension.

5.E The minimal model program with scaling

We start with a $\mathbb{Q}$-factorial klt pair $(X, \Delta + H)$ and a projective morphism $\pi : X \to U$ such that $K_X + \Delta + H$ is nef over U and Δ is big over U. Note that for any klt pair (X, Δ) and any ample divisor H' over U, we may choose $h \in \mathbb{R}_{>0}$ and an $\mathbb{R}$-divisor $H \sim_{\mathbb{R},U} hH'$ such that $(X, \Delta + H)$ is klt and $K_X + \Delta + H$ is nef and big over U. We let

$$\lambda = \inf\{t \geq 0 \mid K_X + \Delta + tH \text{ is nef over } U\}.$$

If $\lambda = 0$, then $K_X + \Delta$ is nef over U and we are done. If $\lambda > 0$, then there exists a $(K_X + \Delta)$-negative extremal ray R over U such that $(K_X + \Delta + \lambda H) \cdot R = 0$ (cf. (5.8)). We consider the corresponding contraction (over U) $\text{cont}_R : X \to Z$. If $\dim Z < \dim X$, then we have a $K_X + \Delta$ Mori fiber space and we are done. Otherwise (assuming the existence of the corresponding flips), we replace (X, Δ) by the corresponding flip or divisorial contraction $\phi : X \dashrightarrow X'$. Let $H' = \phi_* H$ and $\Delta' = \phi_* \Delta$. Since $K_X + \Delta + \lambda H$ is nef over U and $(K_X + \Delta + \lambda H) \cdot R = 0$, it follows that $K_{X'} + \Delta' + \lambda H'$ is nef over U and so we may repeat the process.

As in the usual mmp, this process terminates (yielding an ltm or a Mori fiber space) unless we get an infinite sequence of flips $X_i \dashrightarrow X_{i+1}$. Let Δ_i and H_i denote the strict transforms of Δ and H on X_i. Then there is a sequence of real numbers $\lambda = \lambda_1 \geq \lambda_2 \geq \lambda_3 \geq \cdots > 0$ such that $K_{X_n} + \Delta_n + \lambda_n H_n$ is nef over U. In particular, $X \dashrightarrow X_n$ is an ltm for $(X, \Delta + \lambda_n H)$ over U.

REMARK 5.61. We will need the mmp with scaling in a slightly more general context. We will start with $\pi : X \to U$ a projective morphism and $(X, \Delta + C)$ a dlt $\mathbb{Q}$-factorial pair such that $K_X + \Delta + C$ is nef over U, $\Delta = S + A + B$ where $\lfloor \Delta \rfloor = S$ and $\mathbf{B}_+(A/U)$ contains no non-klt centers of $(X, \Delta + C)$. Notice then that $K_X + \Delta \sim_{\mathbb{R},U} K_X + \Delta'$ where (X, Δ') is klt and Δ' is big. Note that any step of a $(K_X + \Delta)$-mmp over U is also a step of a $(K_X + \Delta')$-mmp over U cf. (5.39). Notice also that the condition that $\mathbf{B}_+(A/U)$ contains no non-klt centers of (X, Δ) is preserved by any $(K_X + \Delta)$-mmp over U cf. (5.43).

Lemma 5.62. *$X_n \dashrightarrow X_m$ is not an isomorphism for any $m > n$.*

Proof. Let E be a divisor over X whose center is contained in $\text{Ex}(\text{cont}_{R_n})$, where $\text{cont}_{R_n} : X_n \to Z_n$ is the corresponding extremal contraction over U. Then

$$a(E; X_n, \Delta_n) < a(E; X_{n+1}, \Delta_{n+1}) \quad \text{and}$$
$$a(E; X_{n+i}, \Delta_{n+i}) \leq a(E; X_{n+i+1}, \Delta_{n+i+1})$$

for all $i > 0$ cf. (5.24) and (5.25). It follows that $a(E; X_n, \Delta_n) < a(E; X_m, \Delta_m)$ and so $X_n \dashrightarrow X_m$ is not an isomorphism. □

Theorem 5.63 (Termination of flips with scaling). *Let $(X, \Delta + H)$ be a $\mathbb{Q}$-factorial klt pair and $\pi : X \to U$ a projective morphism such that $K_X + \Delta + H$ is nef over U and Δ is big over U.*

Then any sequence of flips and divisorial contractions for the $(K_X + \Delta)$-mmp over U with scaling of H is finite.

Proof. By (5.59) and (5.62). □

5.F Pl-flips

Definition 5.64. Let (X, Δ) be a plt pair with $S = \lfloor \Delta \rfloor$. A **pl-flipping contraction** is a flipping contraction $f : X \to Y$ such that $\Delta \in \mathrm{WDiv}_{\mathbb{Q}}(X)$, S is irreducible and $-S$ is f-ample.

REMARK 5.65. By work of V. Shokurov, it is known that the existence of klt flips follows from the existence of pl-flips and special termination cf. [Fuj07].

Definition 5.66. Let (X, Δ) be a plt pair where $\lfloor \Delta \rfloor = S$ is irreducible. The **restricted algebra** $R_S(K_X + \Delta)$ is given by

$$\bigoplus_{m \geq 0} \mathrm{Im}\left(H^0(\mathcal{O}_X(m(K_X + \Delta))) \to H^0(\mathcal{O}_S(m(K_X + \Delta)|_S))\right).$$

We have the following result due to V. Shokurov.

Theorem 5.67. *Let $f : X \to Z$ be a pl-flipping contraction for a plt pair (X, Δ) where $\lfloor \Delta \rfloor = S$. If $Z = \mathrm{Spec}(A)$ is affine, then the flip of f exists if and only if the restricted algebra $R_S(K_X + \Delta)$ is finitely generated.*

Theorem (5.67) is useful because it allows us to proceed by induction on the dimension of X. Before recalling the proof of (5.67), we will need some results about finite generation.

Lemma 5.68. *Let $R = \bigoplus_{m \geq 0} R_m$ be a graded ring and $d > 0$ be an integer. If R is an integral domain, then R is finitely generated if and only if so is $R^{(d)} = \bigoplus_{m \geq 0} R_{md}$.*

Proof. Suppose that R is finitely generated. Since $R^{(d)}$ is the ring of invariants of R by a $\mathbb{Z}/d\mathbb{Z}$ action, then $R^{(d)}$ is finitely generated by a theorem of E. Noether cf. [Eis95, 13.17].

Suppose that $R^{(d)}$ is finitely generated. Note that if $f \in R_m$, then f is integral over $R^{(d)}$ as it is a zero of the monic polynomial $x^d - f^d \in R^{(d)}[x]$. By E. Noether's theorem on the finiteness of integral closures (cf. [Eis95, 13.13]), it follows that R is finitely generated. □

Proof of (5.67). By (5.21), the flip exists if and only if

$$R(K_X + \Delta) = \bigoplus_{m \geq 0} H^0(\mathcal{O}_X(m(K_X + \Delta)))$$

is a finitely generated A-algebra.

If $R(K_X + \Delta)$ is finitely generated, then clearly $R_S(K_X + \Delta)$ is finitely generated.

Assume now that $R_S(K_X + \Delta)$ is finitely generated. Let $S' \sim S$ be an effective divisor not containing S (cf. (5.72)) and let g be a rational function such that $S - S' = (g)$. There are positive integers a and b such that $a(K_X + \Delta) - bS' \equiv_Z 0$. By the Basepoint-free theorem cf. (5.1), $a(K_X + \Delta) - bS'$ is semiample over Z and so we may assume that $a(K_X+\Delta)\sim_Z bS'$. After replacing Z by an open subset, we may assume that $a(K_X+\Delta)\sim bS'$. By (5.68), $R(K_X+\Delta)$ is finitely generated if and only if so is $R(S')$. Since $R_S(K_X+\Delta)$ is finitely generated, by (5.68), $R_S(S')$ is finitely generated and so it suffices to show that the kernel of $\phi : R(S')\to R_S(S')$ is finitely generated. Let g_m be a rational function corresponding to a non-zero element of $R(S')_m$ so that $(g_m)+mS'\geq 0$. If $\phi(g_m)=0$, then $(g_m)+mS'=S+B$ for some $B\geq 0$. But then $(g_m/g)+(m-1)S'=B$ so that g_m/g is a rational function corresponding to a non-zero element of $R(S')_{m-1}$. In particular the kernel of ϕ is the principal ideal generated by g and the result follows. □

One of the main steps of the inductive proof is to show that the existence of log terminal models for $(n-1)$-dimensional pairs of general type (cf. Theorem 5.60) implies the existence of pl-flips for log pairs of dimension n.

Theorem 5.69. *Assume Theorem 5.60 in dimension $n-1$. Let $f : X\to Z$ be a pl-flipping contraction, then the flip $f^+ : X^+\to Z$ exists.*

Proof. See Section 7. □

REMARK 5.70. Many of the ideas in this section are due to V. Shokurov and are explained in [Cor07].

EXERCISE 5.71. Show that if $f: X\to Z$ is a flipping contraction for a dlt pair (X,Δ) such that there is a component S of $\lfloor\Delta\rfloor$ where $-S$ is f-ample, then there is a pair (X,Δ') such that f is a pl-flipping contraction for (X,Δ').

EXERCISE 5.72. Let $f: X\to Z$ be a flipping contraction where Z is affine. Show that for any irreducible divisor $S\subset X$ we have $S\sim S'$ where S' is an effective divisor whose support does not contain S.

5.G Corollaries

Corollary 5.73 (Existence of flips). *Let (X,Δ) be a klt pair and $\pi : X\to U$ be a projective morphism. If $f : X\to Z$ is a $K_X+\Delta$ flipping contraction over U, then the flip f^+ of f exists.*

Proof. Replacing Δ by a sufficiently close $\mathbb{Q}$-divisor, we may assume that $\Delta\in \mathrm{WDiv}_{\mathbb{Q}}(X)$. By (5.22), we may assume that Z is affine. Since Δ is big (over Z), replacing Δ by an appropriate $\mathbb{Q}$-divisor, we may assume that $\Delta=A+B$ where A is an f-ample $\mathbb{Q}$-divisor and $B\geq 0$. By (5.60), $R(K_X+\Delta)$ is finitely generated (over Z) and by (5.21), the flip of f exists and is given by $X^+=\mathrm{Proj}_Z\bigoplus_{m\geq 0} f_*\mathcal{O}_X(m(K_X+\Delta))$. □

Corollary 5.74 (Existence of log terminal models). *Let (X,Δ) be a projective klt pair such that Δ (or $K_X+\Delta$) is big over U and $K_X+\Delta$ is pseudo-effective over U (resp. $K_X+\Delta$ is not pseudo-effective over U).*

Then (X,Δ) has an ltm over U (resp. there is a birational map over U, $X \dashrightarrow \bar{X}$ such that $\bar{X}\to Z$ is a Mori fiber space over U).

Proof. Assume that Δ is big over U. Let $f : Y \to X$ be a log resolution and write

$$K_Y + \Gamma = f^*(K_X + \Delta) + E$$

as in (3.21.1). We may assume that there is an effective divisor F whose support equals the exceptional set. For $0 < \varepsilon \ll 1$, $K_Y + \Gamma + \varepsilon F$ is klt and $\Gamma + \varepsilon F$ is big. By (5.54) there is an ltm $g : Y \dashrightarrow Y'$ for $K_Y + \Gamma + \varepsilon F$ over U. It suffices to show that $\phi = g \circ f^{-1}$ is an ltm for $K_X + \Delta$ over U.

Let $p : W \to X, q : W \to Y$ and $r : W \to Y'$ be a common log resolution. Then

$$p^*(K_X + \Delta) + q^*(E + \varepsilon F) = q^*(K_Y + \Gamma + \varepsilon F) = r^*(K_{Y'} + g_*(\Gamma + \varepsilon F)) + G$$

where G is effective and r-exceptional. Since $K_{Y'} + g_*(\Gamma + \varepsilon F)$ is semiample over U cf. (5.10), it follows that $\mathbf{B}(r^*(K_{Y'} + g_*(\Gamma + \varepsilon F)) + G/U) = \operatorname{Supp}(G)$. Since $q^*(E + \varepsilon F)$ is q-exceptional, we have that

$$\operatorname{Supp}(q^*(E + \varepsilon F)) \subset \mathbf{B}(p^*(K_X + \Delta) + q^*(E + \varepsilon F)/U) = \operatorname{Supp}(G).$$

Therefore, $\phi : X \dashrightarrow Y'$ extracts no divisors (i.e., $Y' \dashrightarrow X$ contracts no divisors) and $g_*F = 0$. For any ϕ-exceptional divisor $P \subset X$, we have

$$\begin{aligned} a(P; X, \Delta) &= a(P; Y, \Gamma) = a(P; Y, \Gamma + \varepsilon F) \\ &< a(P; Y', g_*(\Gamma + \varepsilon F)) = a(P; Y', g_*\Gamma) = a(P; Y', \phi_*\Delta). \end{aligned}$$

Therefore $X \dashrightarrow Y$ is an ltm for (X, Δ).

The case when $K_X + \Delta$ is big over U is similar. By (3.35), there is a klt pair (X, Δ') such that $K_X + \Delta' \sim_{\mathbb{R},U} (1+\varepsilon)(K_X + \Delta)$ for some $\varepsilon > 0$. Since $\Delta' \sim_{\mathbb{R},U} \Delta + \varepsilon(K_X + \Delta)$, Δ' is big over U. By what we have already shown, there is an ltm for (X, Δ') over U. By (5.40), an ltm for (X, Δ') over U is an ltm for (X, Δ) over U.

We may therefore assume that $K_X + \Delta$ is not pseudo-effective over U. Let A be an ample $\mathbb{Q}$-divisor and $t > 0$ such that $K_X + \Delta + A$ is nef over U and klt. Let

$$\tau = \inf\{t > 0 \mid K_X + \Delta + tA \text{ is big}\}.$$

Since $K_X + \Delta$ is not pseudo-effective over U, $\tau > 0$. By what we have seen above, there exists an ltm $\phi : X \dashrightarrow X'$ for $(X, \Delta + \tau A)$ over X. In particular X' is $\mathbb{Q}$-factorial and $X' \to X$ is a small birational morphism. If Δ' and A' denote the strict transforms of Δ and A, then $K_{X'} + \Delta' + \tau A'$ is nef over U but not big over U. By (5.54) there is a finite sequence of flips and divisorial contractions over U, $X' \dashrightarrow X'_n$ such that $X'_n \to Z$ is a Mori fiber space over U. □

Corollary 5.75 (Finite generation). *Let X be a smooth projective variety. Then $R(K_X)$ is finitely generated.*

Proof. By [FM00], there exists a projective klt pair (Z, Δ_Z) such that $K_Z + \Delta_Z$ is big and $R(K_X)$ is finitely generated if and only if so is $R(K_Z + \Delta_Z)$. The statement now follows from (5.74), (5.11) and (5.10). □

Corollary 5.76 (Finiteness of models). *Assume Theorems 5.57, 5.58, 5.59 in dimensions $\leq n$. Let $\pi : S \to U$ be a projective morphism of normal quasi-projective varieties, $(S, \Delta + C)$ an n-dimensional klt pair and $S \dashrightarrow S_i$ an infinite sequence of small birational maps of normal varieties over U. Let Δ_i and C_i be the strict transforms of Δ and C. If Δ is big over U and there are numbers $0 \leq \lambda_i \leq 1$ such that $K_{S_i} + \Delta_i + \lambda_i C_i$ is $\mathbb{R}$-Cartier and nef over U, then the set of rational maps $\phi_i : S \dashrightarrow S_i$ is finite modulo isomorphisms over U (i.e., there exist finitely many rational maps $\psi_j : S \dashrightarrow T_j$ over U such that for any i, there exists an integer j such that $\phi_i \circ \psi_j^{-1} : T_j \dashrightarrow S_i$ is an isomorphism).*

Proof. Replacing S by an ltm for $K_S + \Delta + C$ over S, we may assume that $(S, \Delta + C)$ is $\mathbb{Q}$-factorial. Since $(S, \Delta + \lambda_i C)$ is klt, by (5.23), it follows that $(S_i, \Delta_i + \lambda_i C_i)$ is klt. Let $S_i \dashrightarrow S_i'$ be an ltm for $K_{S_i} + \Delta_i + \lambda_i C_i$ over S_i and let $\nu_i : S_i' \to S_i$ be the induced morphism. Then $\eta_i : S \dashrightarrow S_i'$ is a small birational morphism and so $K_{S_i'} + \Delta_i' + \lambda_i C_i' = \nu_i^*(K_{S_i} + \Delta_i + \lambda_i C_i)$ where Δ_i' and C_i' are the strict transforms of Δ_i and C_i. It follows that η_i is an ltm of $(S, \Delta + \lambda_i C)$ over U. By (5.59) in dimension n, the set of rational maps $\eta_i : S \dashrightarrow S_i'$ is finite modulo isomorphisms over U. The map $\nu_i : S_i' \to S_i$ is given by the choice of a $(K_{S_i'} + \Delta_i' + \lambda_i C_i')$-trivial extremal face of $\overline{\mathrm{NE}}(S_i'/U)$. Since Δ is big over U, $\Delta_i' + \lambda_i C_i' \sim_{\mathbb{R},U} A + B$ where A is ample over U and $(S_i', A + B)$ is klt. The above face is $(K_{S_i'} + \frac{1}{2}A + B)$-negative. By the Cone theorem (5.4), there are finitely many $(K_{S_i'} + \frac{1}{2}A + B)$-negative extremal rays over U. Therefore, for any i, there are finitely many morphisms $S_i' \to S_i$. □

Corollary 5.77 (Non-vanishing). *Assume Theorems 5.57, 5.58, 5.59 in dimensions $\leq n$. Let (X, Δ) be a projective $\mathbb{Q}$-factorial klt pair where Δ is big and $K_X + \Delta$ is pseudo-effective. Then there exist $r_i \in \mathbb{R}_{\geq 0}$ and $\Delta_i \in \mathrm{WDiv}_{\mathbb{Q}}(X)$ such that*

(5.77.1) $\mathrm{Supp}(\Delta_i) = \mathrm{Supp}(\Delta)$,

(5.77.2) $K_X + \Delta = \sum r_i(K_X + \Delta_i)$ *and*

(5.77.3) $|K_X + \Delta_i|_{\mathbb{Q}} \neq 0$.

Proof. By (5.58) and (5.57) there is an ltm $\phi : X \dashrightarrow Y$ for $K_X + \Delta$. Let $V \subset \mathrm{WDiv}_{\mathbb{Q}}(X)$ be the smallest rational affine subspace containing Δ and pick $r_i \in \mathbb{R}$ and $\Delta_i \in V$ such that $K_X + \Delta = \sum r_i(K_X + \Delta_i)$. Notice that $\mathrm{Supp}(\Delta_i) = \mathrm{Supp}(\Delta)$. We claim that if $\|\Delta - \Delta_i\| \ll 1$, then ϕ is an ltm for $K_X + \Delta_i$ for all i. Since $a(E; X, \Delta) < a(E; Y, \phi_*\Delta)$, by continuity it follows that $a(E; X, \Delta_i) < a(E; Y, \phi_*\Delta_i)$. It suffices then to show that $K_Y + \phi_*\Delta_i$ is nef for all i. Since $\phi_*\Delta$ is big, we may write $\phi_*\Delta \sim_{\mathbb{R}} A + B$ where A is an ample $\mathbb{Q}$-divisor and $(Y, A + B)$ is klt. We may assume that $A/2 + \phi_*\Delta_i - \phi_*\Delta$ is ample for all i. If $K_Y + \phi_*\Delta_i$ is not nef, then there is an extremal ray $R = \mathbb{R}_{\geq 0}[C]$ such that $(K_Y + \phi_*\Delta_i) \cdot C < 0$. But then

$$\left(K_Y + \frac{A}{2} + B\right) \cdot C = (K_Y + \phi_*\Delta_i) \cdot C - \left(\frac{A}{2} + \phi_*\Delta_i - \phi_*\Delta\right) \cdot C < 0.$$

Therefore R is a negative extremal ray for $K_Y + A/2 + B$. By (5.4), there are only finitely many such rays and so

$$\{\Delta' \in V \mid K_Y + \phi_* \Delta' \text{ is nef}\}$$

is a rational polyhedron containing Δ in its interior. It follows that $K_Y + \phi_* \Delta_i$ is nef for all i. By (5.10), $|K_Y + \phi_* \Delta_i|_{\mathbb{Q}} \neq \emptyset$ and so $|K_X + \Delta_i|_{\mathbb{Q}} \neq \emptyset$. □

CHAPTER 6

Multiplier ideal sheaves

In this chapter we will recall the definition and the main properties of multiplier ideal sheaves. The standard reference for multiplier ideal sheaves is [Laz04b]. In what follows we will focus on a generalization of this notion known as adjoint ideals.

Definition 6.1. Let X be a smooth quasi-projective variety, Δ be a reduced divisor with simple normal crossings support and V a linear series such that $\mathrm{Bs}(V)$ contains no non-klt center of (X, Δ) (i.e., $\mathrm{Bs}(V)$ contains no strata of Δ). Let $f: Y \to X$ be a log resolution of V and of (X, Δ). We let

$$K_Y + \Gamma = f^*(K_X + \Delta) + E$$

as in (3.21.1).

For any real number $c > 0$, we define the **multiplier ideal sheaf**

$$\mathcal{J}_{\Delta, c \cdot V} := f_* \mathcal{O}_Y(E - \lfloor cF \rfloor)$$

where $F = \mathrm{Fix}(f^* V)$.

If $D \geq 0$ is a $\mathbb{Q}$-divisor and $m > 0$ is an integer such that mD is Cartier, then we let $\mathcal{J}_{\Delta, cD} = \mathcal{J}_{\Delta, \frac{c}{m} \cdot V}$ where $V = \{mD\}$ is the linear series consisting of the unique divisor mD.

REMARK 6.2. If $\Delta = 0$, then $\mathcal{J}_{\Delta, c \cdot V} = \mathcal{J}_{c \cdot V}$ is the usual multiplier ideal sheaf.

Lemma 6.3. *The definition given in* (6.1) *is independent of the log resolution.*

Proof. Let $f: Y \to X$ and $f': Y' \to X$ be two log resolutions. Since any two log resolutions can be dominated by a third log resolution, we may assume that there is a morphism $\mu: Y' \to Y$ such that $f' = f \circ \mu$. We let

$$K_Y + \Gamma = f^*(K_X + \Delta) + E \qquad \text{and} \qquad K_{Y'} + \Gamma' = (f')^*(K_X + \Delta) + E'$$

be as in (3.21.1). We have

$$K_{Y'} + \Gamma' = \mu^*(K_Y + \Gamma - E) + E'.$$

Let $F = \mathrm{Fix}(f^*V)$ and $F' = \mathrm{Fix}((f')^*V)$. Since $f^*V - F$ is basepoint-free, it follows that $F' = \mu^*F$. We claim that

$$\mu_*\mathcal{O}_{Y'}(E' - \mu^*E - \lfloor \mu^*\{cF\}\rfloor) = \mathcal{O}_Y.$$

Grant this for the time being, then

$$\begin{aligned}(f')_*\mathcal{O}_{Y'}(E' - \lfloor cF' \rfloor) &= (f')_*\mathcal{O}_{Y'}(E' - \mu^*E - \lfloor \mu^*\{cF\}\rfloor + \mu^*(E - \lfloor cF \rfloor)) \\ &= f_*\left(\mu_*\mathcal{O}_{Y'}(E' - \mu^*E - \lfloor \mu^*\{cF\}\rfloor) \otimes \mathcal{O}_Y(E - \lfloor cF \rfloor)\right) = f_*\mathcal{O}_Y(E - \lfloor cF \rfloor).\end{aligned}$$

The statement now follows immediately. To see the claim, notice that $K_Y + \Gamma + \{cF\}$ is dlt and its non-klt places coincide with those of $K_Y + \Gamma - E$. From the equation

$$K_{Y'} + \Gamma' + \mu^*(E + \{cF\}) - E' = \mu^*(K_Y + \Gamma + \{cF\})$$

it follows that $\lfloor \mu^*(E + \{cF\}) - E' \rfloor \leq 0$ so that $E' - \mu^*E - \lfloor \mu^*\{cF\}\rfloor$ is effective and μ-exceptional. The claim is now clear. □

Lemma 6.4. *Let X be a smooth quasi-projective variety, Δ be a reduced divisor with simple normal crossings support and $V \subset V'$ be linear series such that* Bs(V) *contains no non-klt center of (X, Δ). Let $c \geq c'$ be positive real numbers, $\Delta' \leq \Delta$ be reduced divisors and D and D' be effective $\mathbb{Q}$-Cartier divisors whose supports contain no non-klt centers of (X, Δ). Then:*

(6.4.1) $\mathcal{J}_{\Delta, c\cdot V} \subset \mathcal{J}_{\Delta', c'\cdot V'}$, and in particular $\mathcal{J}_{\Delta, c\cdot V} \subset \mathcal{J}_{c\cdot V} \subset \mathcal{O}_X$.

(6.4.2) If Σ is an effective Cartier divisor such that $D \leq \Sigma + D'$ and $\mathcal{J}_{\Delta, D'} = \mathcal{O}_X$, then $\mathcal{I}_\Sigma \subset \mathcal{J}_{\Delta, D}$.

Proof. Let $f: Y \to X$ be a common log resolution of V, V', (X, Δ) and (X', Δ'). We let

$$K_Y + \Gamma = f^*(K_X + \Delta) + E \qquad \text{and} \qquad K_Y + \Gamma' = f^*(K_X + \Delta') + E'$$

be as in (3.21.1). We have that $\Gamma' \leq \Gamma$ and $E' \geq E$. If we let $F = \mathrm{Fix}(f^*V)$ and $F' = \mathrm{Fix}(f^*V')$, then we also have $F \geq F'$. It follows that

$$\mathcal{J}_{\Delta, c\cdot V} = f_*\mathcal{O}_Y(E - \lfloor cF \rfloor) \subset f_*\mathcal{O}_Y(E' - \lfloor c'F' \rfloor) = \mathcal{J}_{\Delta', c'\cdot V'}.$$

Let $f: Y \to X$ be a log resolution of $(X, \Delta + \Sigma + D')$. We let

$$K_Y + \Gamma = f^*(K_X + \Delta) + E$$

be as in (3.21.1). Since $\mathcal{J}_{\Delta, D'} = \mathcal{O}_X$, we have $E - \lfloor f^*D' \rfloor \geq 0$. Since Σ is Cartier, we have $\lfloor f^*D \rfloor \leq f^*\Sigma + \lfloor f^*D' \rfloor$. Therefore

$$E - \lfloor f^*D \rfloor \geq E - \lfloor f^*D' \rfloor - f^*\Sigma \geq -f^*\Sigma$$

and hence

$$\mathcal{I}_\Sigma = f_*\mathcal{O}_Y(-f^*\Sigma) \subset f_*\mathcal{O}_Y(E - \lfloor f^*D \rfloor) = \mathcal{J}_{\Delta, D}.$$

□

The next lemma shows that multiplier ideals of the form $\mathcal{J}_{\Delta,D}$ may be viewed as being obtained by successive extensions of the usual multiplier ideal sheaves $\mathcal{J}_D$.

Lemma 6.5. *Let X be a smooth quasi-projective variety, Δ be a reduced divisor with simple normal crossings support and $D \geq 0$ be a $\mathbb{Q}$-Cartier divisor whose support contains no non-klt center of (X, Δ). Let S be a component of Δ, then there is a short exact sequence*

$$0 \to \mathcal{J}_{\Delta-S,D+S} \to \mathcal{J}_{\Delta,D} \to \mathcal{J}_{(\Delta-S)|_S,D|_S} \to 0.$$

Proof. Let $f : Y \to X$ be a log resolution of $(X, \Delta + D)$ which is an isomorphism at the generic point of each non-klt center of (X, Δ) cf. (3.22). We write

$$K_Y + \Gamma = f^*(K_X + \Delta) + E$$

as in (3.21.1). Let $T = f_*^{-1}S$ and consider the short exact sequence

$$0 \to \mathcal{O}_Y(E - \lfloor f^*D \rfloor - T) \to \mathcal{O}_Y(E - \lfloor f^*D \rfloor) \to \mathcal{O}_T(E - \lfloor f^*D \rfloor) \to 0.$$

By definition, we have $f_*\mathcal{O}_Y(E - \lfloor f^*D \rfloor) = \mathcal{J}_{\Delta,D}$. Since

$$K_Y + \Gamma - T = f^*(K_X + \Delta - S) + E + f^*S - T$$

where $\Gamma - T$ and $E + f^*S - T$ are effective with no common components, it follows that

$$\mathcal{J}_{\Delta-S,D+S} = f_*\mathcal{O}_Y(E + f^*S - T - \lfloor f^*(D + S) \rfloor) = f_*\mathcal{O}_Y(E - \lfloor f^*D \rfloor - T).$$

Since

$$K_T + (\Gamma - T)|_T = (f|_T)^*(K_S + (\Delta - S)|_S) + E|_T,$$

it follows that

$$\mathcal{J}_{(\Delta-S)|_S,D|_S} = (f|_T)_*\mathcal{O}_T(E - \lfloor f^*D \rfloor).$$

Finally, pick an effective and f-exceptional $\mathbb{Q}$-divisor F such that $-F$ is f-ample and $(Y, \Gamma - T + \{f^*D\} + F)$ is dlt cf. (6.24). Since

$$E - \lfloor f^*D \rfloor - T = K_Y + \Gamma - T + \{f^*D\} + F - F - f^*(K_X + \Delta + D),$$

by Kawamata-Viehweg vanishing (cf. (3.46)), we have that $\mathcal{R}^1 f_*\mathcal{O}_Y(E - \lfloor f^*D \rfloor - T) = 0$. The lemma now follows by pushing forward the above short exact sequence. □

Lemma 6.6. *Let X be a smooth quasi-projective variety, Δ be a reduced divisor with simple normal crossings support and $D \geq 0$ be a $\mathbb{Q}$-Cartier divisor whose support contains no non-klt center of (X, Δ). Let $\pi: X \to Z$ be a projective morphism to a normal affine variety Z and N be a Cartier divisor on X such that $N - D$ is ample. Then*

$$H^i(X, \mathcal{J}_{\Delta,D}(K_X + \Delta + N)) = 0 \qquad \text{for } i > 0,$$

and if S is a component of Δ, then the homomorphism

$$H^0(X, \mathcal{J}_{\Delta,D}(K_X + \Delta + N)) \to H^0(S, \mathcal{J}_{(\Delta-S)|_S,D|_S}(K_X + \Delta + N))$$

is surjective.

Proof. Let $f : Y \to X$ be a log resolution of $(X, \Delta + D)$ which is an isomorphism at the generic point of each non-klt center of (X, Δ) cf. (3.22). We write

$$K_Y + \Gamma = f^*(K_X + \Delta) + E$$

as in (3.21.1). Pick an effective and f-exceptional $\mathbb{Q}$-divisor F such that $f^*(N - D) - F$ is ample and $(Y, \Gamma + \{f^*D\} + F)$ is dlt. Since

$$E - \lfloor f^*D \rfloor + f^*(K_X + \Delta + N) = K_Y + \Gamma + \{f^*D\} + F + f^*(N - D) - F,$$

by Kawamata-Viehweg vanishing (3.46), we have that $\mathcal{R}^j f_*\mathcal{O}_Y(E - \lfloor f^*D \rfloor) = 0$ for all $j > 0$. Since $f_*\mathcal{O}_Y(E - \lfloor f^*D \rfloor) = \mathcal{J}_{\Delta,D}$, by Kawamata-Viehweg vanishing (3.46) and an easy spectral sequence argument (6.25), it follows then that

$$H^i(X, \mathcal{J}_{\Delta,D}(K_X + \Delta + N)) = H^i(Y, \mathcal{O}_Y(E - \lfloor f^*D \rfloor + f^*(K_X + \Delta + N))) = 0$$

for all $i > 0$.

It also follows that $H^i(X, \mathcal{J}_{\Delta-S,D+S}(K_X + \Delta + N)) = 0$ for all $i > 0$ and hence the final claim follows from (6.5). □

6.A Asymptotic multiplier ideal sheaves

Definition 6.7. Let D be a divisor on a normal variety X and $V_i \subset |iD|$ be linear series, then $V_\bullet$ is an **additive sequence of linear series** if

$$V_i + V_j \subset V_{i+j}$$

for all $i, j \geq 0$.

Lemma 6.8. *Let X be a smooth quasi-projective variety and Δ be a reduced divisor with simple normal crossings support. If $V_\bullet$ is an additive sequence of linear series and c is a positive real number, k a positive integer such that* $\mathrm{Bs}(V_k)$ *contains no non-klt centers of (X, Δ), then*

$$\mathcal{J}_{\Delta,\frac{c}{p}\cdot V_p} \subset \mathcal{J}_{\Delta,\frac{c}{q}\cdot V_q}$$

for any positive integers p and q such that k divides p and p divides q.

In particular, since X is Noetherian, the set $\{\mathcal{J}_{\Delta,\frac{c}{p}\cdot V_p}\}$ has a unique maximal element $\mathcal{J}_{\Delta,\|c\cdot V_\bullet\|}$. Note that $\mathcal{J}_{\Delta,\|c\cdot V_\bullet\|} = \mathcal{J}_{\Delta,\frac{c}{p}\cdot V_p}$ for any $p > 0$ sufficiently divisible.

Proof. This follows from (1) of (6.4), since $(q/p)V_p \subset V_q$. □

Definition 6.9. Let X be a smooth quasi-projective variety, Δ be a reduced divisor with simple normal crossings support and S an irreducible component of Δ. If c is a positive real number, D is a $\mathbb{Q}$-divisor on X such that $\mathbf{B}(D)$ contains no non-klt centers of (X, Δ), then we define the **asymptotic multiplier ideal sheaves**

$$\mathcal{J}_{\Delta,c\cdot\|D\|} = \mathcal{J}_{\Delta,\|c\cdot V_\bullet\|} \quad \text{and} \quad \mathcal{J}_{(\Delta-S)|_S,c\cdot\|D\|_S} = \mathcal{J}_{(\Delta-S)|_S,\|c\cdot W_\bullet\|}$$

where $V_m = |mD|$ and $W_m = |mD|_S$ is the restriction (or trace) of the linear series $|mD|$ to S.

Lemma 6.10. *Let X be a smooth quasi-projective variety, Δ be a reduced divisor with simple normal crossings support, S an irreducible component of Δ and D a $\mathbb{Q}$-divisor on X such that $\mathbf{B}(D)$ contains no non-klt centers of (X, Δ). Let $\pi: X \to Z$ be a projective morphism to a normal affine variety. Then:*

(6.10.1) For any real numbers $0 < c' \leq c$ and any reduced divisor $0 \leq \Delta' \leq \Delta$, we have

$$\mathscr{J}_{\Delta,\|cD\|} \subset \mathscr{J}_{\Delta',\|c'D\|}.$$

(6.10.2) If D is Cartier, then

$$\mathrm{Im}\,(\pi_*\mathscr{O}_X(D) \to \pi_*\mathscr{O}_S(D)) \subset \pi_*\mathscr{J}_{(\Delta-S)|_S,\|D\|_S}(D).$$

(6.10.3) If D is Cartier, A is an ample Cartier divisor and H is a very ample Cartier divisor, then

$$\mathscr{J}_{\|D\|}(K_X + D + A + nH)$$

is globally generated where $n = \dim X$.

(6.10.4) If D is Cartier and $\mathbf{B}_+(D)$ contains no non-klt centers of (X, Δ), then the image of the homomorphism

$$\pi_*\mathscr{O}_X(K_X + \Delta + D) \to \pi_*\mathscr{O}_S(K_X + \Delta + D)$$

contains

$$\pi_*\mathscr{J}_{(\Delta-S)|_S,\|D\|_S}(K_X + \Delta + D).$$

Proof. (1) is immediate from (1) of (6.4).

To see (2), let $p > 0$ be an integer such that $\mathscr{J}_{(\Delta-S)|_S,\|D\|_S} = \mathscr{J}_{(\Delta-S)|_S,\frac{1}{p}\cdot|pD|_S}$ and consider a log resolution $f: Y \to X$ of (X, Δ) and $|pD|$. We let

$$K_Y + \Gamma = f^*(K_X + \Delta) + E$$

be as in (3.21.1). Let $F_p = \mathrm{Fix}(f^*|pD|)$, then $pF_1 \geq F_p$ so that

$$\begin{aligned}\pi_*\mathscr{O}_X(D) &= (\pi \circ f)_*\mathscr{O}_Y(f^*D - F_1)\\ &\subseteq (\pi \circ f)_*\mathscr{O}_Y\left(f^*D + E - \left\lfloor \frac{1}{p}F_p \right\rfloor\right)\\ &\subseteq (\pi \circ f)_*\mathscr{O}_Y(f^*D + E) = \pi_*\mathscr{O}_X(D).\end{aligned}$$

The assertion now follows as

$$\pi_*\mathscr{J}_{(\Delta-S)|_S,\|D\|_S}(D) = (\pi \circ f)_*\mathscr{O}_T(f^*D + E - \left\lfloor \frac{1}{p}F_p \right\rfloor)$$

where $T = f_*^{-1}S$.

(3) is immediate from (6.6) and (6.26).

To see (4), let $p > 0$ be an integer such that $pD \sim A + B$ where A is a general very ample divisor and B is an effective divisor containing no non-klt centers of (X, Δ).

Replacing p by a multiple, we may assume that $\mathscr{J}_{(\Delta-S)|_S,\|D\|_S} = \mathscr{J}_{(\Delta-S)|_S,\frac{1}{p}\cdot|pD|_S}$. Let $f\colon Y \to X$ be a log resolution of (X,Δ) and $|pD|$, which is an isomorphism at the generic point of each non-klt center of (X,Δ) cf. (3.22). We let

$$K_Y + \Gamma = f^*(K_X + \Delta) + E$$

be as in (3.21.1). Let $F_p = \mathrm{Fix}(f^*|pD|)$, then $M_p := pf^*D - F_p$ is basepoint-free and $f^*B \geq F_p$. Let $T = f_*^{-1}S$ and F be an effective and exceptional divisor such that $f^*A - F$ is ample and $\delta > 0$ be a rational number such that $K_Y + \Gamma - T + \{\frac{1}{p}F_p\} + \delta(f^*B - F_p + F)$ is dlt. We have

$$\begin{aligned} E - \left\lfloor \frac{1}{p}F_p \right\rfloor - T + f^*(K_X + \Delta + D) &= K_Y + \Gamma - T + f^*D - \left\lfloor \frac{1}{p}F_p \right\rfloor \\ \sim_{\mathbb{Q}} K_Y + \Gamma - T + \left\{ \frac{1}{p}F_p \right\} + \delta(f^*B - F_p + F) &+ \left(\frac{1}{p} - \delta \right) M_p + \delta(f^*A - F). \end{aligned}$$

By Kawamata-Viehweg vanishing (3.46),

$$\mathcal{R}^1(\pi \circ f)_*\mathscr{O}_Y\left(E - \left\lfloor \frac{1}{p}F_p \right\rfloor - T + f^*(K_X + \Delta + D)\right) = 0.$$

Therefore, pushing forward the short exact sequence

$$\begin{aligned} 0 \to \mathscr{O}_Y\left(E - \left\lfloor \frac{1}{p}F_p \right\rfloor - T + f^*(K_X + \Delta + D)\right) & \\ \to \mathscr{O}_Y\left(E - \left\lfloor \frac{1}{p}F_p \right\rfloor + f^*(K_X + \Delta + D)\right) & \\ \to \mathscr{O}_T\left(E - \left\lfloor \frac{1}{p}F_p \right\rfloor + f^*(K_X + \Delta + D)\right) \to 0, & \end{aligned}$$

one sees that the homomorphism

$$\begin{aligned} (\pi \circ f)_*\mathscr{O}_Y\left(E - \left\lfloor \frac{1}{p}F_p \right\rfloor + f^*(K_X + \Delta + D)\right) & \\ \to (\pi \circ f)_*\mathscr{O}_T\left(E - \left\lfloor \frac{1}{p}F_p \right\rfloor + f^*(K_X + \Delta + D)\right) & \end{aligned}$$

is surjective. (4) now follows as

$$(\pi \circ f)_*\mathscr{O}_Y\left(E - \left\lfloor \frac{1}{p}F_p \right\rfloor + f^*(K_X + \Delta + D)\right) \subseteq \pi_*\mathscr{O}_X(K_X + \Delta + D)$$

and

$$(\pi \circ f)_*\mathscr{O}_T\left(E - \left\lfloor \frac{1}{p}F_p \right\rfloor + f^*(K_X + \Delta + D)\right) = \pi_*\mathscr{J}_{(\Delta-S)|_S,\|D\|_S}(K_X + \Delta + D).$$

$\square$

6.B Extending pluricanonical forms

Theorem 6.11. *Let $\pi: X \to Y$ be a projective morphism from a smooth quasi-projective variety to an affine variety, $\Delta = \sum \delta_i \Delta_i$ be a divisor with simple normal crossings support and rational coefficients $0 \le \delta_i \le 1$ and S an irreducible component of $\lfloor \Delta \rfloor$. Let $k > 0$ be an integer such that $D = k(K_X + \Delta)$ is Cartier. If $\mathbf{B}(D)$ contains no non-klt centers of $(X, \lceil \Delta \rceil)$ and if A is a sufficiently ample Cartier divisor, then for all $m > 0$,*

$$(\star_m) \qquad \mathcal{J}_{\|mD|_S\|} \subseteq \mathcal{J}_{(\lceil\Delta\rceil - S)|_S, \|mD+A\|_S}, \qquad \textit{and}$$
$$\pi_* \mathcal{J}_{\|mD|_S\|}(mD + A) \subseteq \operatorname{Im}\left(\pi_* \mathcal{O}_X(mD + A) \to \pi_* \mathcal{O}_S(mD + A)\right).$$

Proof. We proceed by induction on m. Since $\mathcal{J}_{(\lceil\Delta\rceil - S)|_S, \|A\|_S} = \mathcal{O}_S$, $(\star_0)$ is clear. Therefore we must show that $(\star_{m+1})$ holds assuming that $(\star_m)$ holds. It is easy to see that there is a unique choice of integral reduced divisors

$$S \le \Delta^1 \le \Delta^2 \le \cdots \le \Delta^k = \Delta^{k+1} = \lceil \Delta \rceil$$

such that $k\Delta = \sum_{i=1}^k \Delta^i$ (cf. (6.27)). We let

$$D_{\le 0} = 0 \qquad \text{and} \qquad D_{\le s} = K_X + \Delta^s + D_{\le s-1} \ \text{ for } 0 < s \le k+1.$$

In particular $D = D_{\le k}$. Since $\mathcal{J}_{\|(m+1)D|_S\|} \subset \mathcal{J}_{\|mD|_S\|}$ (cf. (1) of (6.10)) and since $\mathcal{J}_{(\Delta^{k+1} - S)|_S, \|mD + D_{\le k} + A\|_S} = \mathcal{J}_{(\lceil\Delta\rceil - S)|_S, \|(m+1)D + A\|_S}$, to prove $(\star_{m+1})$, it suffices to show that

$$(\sharp_i) \qquad \mathcal{J}_{\|mD|_S\|} \subseteq \mathcal{J}_{(\Delta^{i+1} - S)|_S, \|mD + D_{\le i} + A\|_S} \qquad \text{for } 0 \le i \le k.$$

We proceed by induction on $0 \le i \le k$. $(\sharp_0)$ holds since by $(\star_m)$ and (1) of (6.10), we have

$$\mathcal{J}_{\|mD|_S\|} \subseteq \mathcal{J}_{(\lceil\Delta\rceil - S)|_S, \|mD+A\|_S} \subseteq \mathcal{J}_{(\Delta^1 - S)|_S, \|mD+A\|_S}.$$

Assume that $(\sharp_{i-1})$ holds for some $0 < i \le k$. We have that

$$\begin{aligned}
\pi_* \mathcal{J}_{\|mD|_S\|}(mD + D_{\le i} + A) &\subseteq \pi_* \mathcal{J}_{(\Delta^i - S)|_S, \|mD + D_{\le i-1} + A\|_S}(mD + D_{\le i} + A) \\
&\subseteq \operatorname{Im}\left(\pi_* \mathcal{O}_X(mD + D_{\le i} + A) \to \pi_* \mathcal{O}_S(mD + D_{\le i} + A)\right) \\
&\subseteq \pi_* \mathcal{J}_{(\Delta^{i+1} - S)|_S, \|mD + D_{\le i} + A\|_S}(mD + D_{\le i} + A).
\end{aligned}$$

The first inclusion follows as we are assuming $(\sharp_{i-1})$. The second inclusion follows by (4) of (6.10), since we may assume that $D_{\le i-1} + A$ is ample and hence that $\mathbf{B}_+(mD + D_{\le i-1} + A)$ contains no non-klt centers of (X, Δ^i). The third inclusion follows from (2) of (6.10) since we may assume that $D_{\le i} + A$ is ample and hence that $\mathbf{B}(mD + D_{\le i} + A)$ contains no non-klt centers of (X, Δ^{i+1}).

By (3) of (6.10)

$$\mathcal{J}_{\|mD|_S\|}(mD + D_{\le i} + A) = \mathcal{J}_{\|mD|_S\|}(K_S + mD|_S + (\Delta^i - S + D_{\le i-1} + A)|_S)$$

is generated by global sections since $(\Delta^i - S + D_{\le i-1} + A)|_S$ is sufficiently ample. Therefore, $(\sharp_i)$ holds. This concludes the proof of $(\star_m)$ for all $m \ge 0$. The remaining claim is clear from what we have shown above. □

Theorem 6.12. *Let $\pi\colon X \to Z$ be a projective morphism from a smooth quasi-projective variety to an affine variety. Let $\Delta = S + A + B$ be a $\mathbb{Q}$-divisor such that $S = \lfloor \Delta \rfloor$ is irreducible and smooth, $A \geq 0$ is a general ample $\mathbb{Q}$-divisor, $B \geq 0$, (X, Δ) is plt, $(S, \Omega + A|_S)$ is canonical where $\Omega = (A + B)|_S$, and $\mathbf{B}(K_X + \Delta)$ does not contain S. For any sufficiently divisible integer $m > 0$ let*

$$F_m = \operatorname{Fix}(|m(K_X + \Delta)|_S)/m \qquad \text{and} \qquad F = \lim F_{m!}.$$

Let $\varepsilon > 0$ be a rational number such that $\varepsilon(K_X + \Delta) + A$ is ample. If Φ is a $\mathbb{Q}$-divisor on S and $k > 0$ is an integer such that $k\Delta$ and $k\Phi$ are Cartier and

$$\Omega \wedge \left(1 - \frac{\varepsilon}{k}\right) F \leq \Phi \leq \Omega,$$

then

$$|k(K_S + \Omega - \Phi)| + k\Phi \subseteq |k(K_X + \Delta)|_S.$$

REMARK 6.13. Since $(m+1)! = (m+1)m!$, we have $F_{(m+1)!} \leq F_{m!}$ and so $\lim F_{m!} = \inf\{F_{m!}\}$ exists.

Proof. Since A is a general ample $\mathbb{Q}$-divisor, $\frac{k-1}{k}A$ is a multiple of a general very ample divisor and so $(X, \Delta + \frac{k-1}{k}A)$ is plt. By assumption $(S, \Omega + \frac{1}{k}A|_S)$ is canonical. Let $l > 0$ be a sufficiently big and divisible integer.

We will show (see (‡) below) that there exists an effective divisor H on X whose support does not contain S such that for all integers $m > 0$ divisible by l, we have

$$(\dagger) \qquad |m(K_S + \Omega - \Phi)| + m\Phi + \left(\frac{m}{k}A + H\right)\Big|_S \subset |m(K_X + \Delta) + \frac{m}{k}A + H|_S.$$

Grant this for the time being. Then, for any $\Sigma \in |k(K_S + \Omega - \Phi)|$, we may choose a divisor $G \in |m(K_X + \Delta) + \frac{m}{k}A + H|$ such that $G|_S = \frac{m}{k}\Sigma + m\Phi + (\frac{m}{k}A + H)|_S$. If we define $\Lambda = \frac{k-1}{m}G + B$, then

$$k(K_X + \Delta) \sim_{\mathbb{Q}} K_X + S + \Lambda + \left(\frac{1}{k}A - \frac{k-1}{m}H\right)$$

where $\frac{1}{k}A - \frac{k-1}{m}H$ is ample as m is sufficiently big. By (6.6), we have a surjective homomorphism

$$H^0(X, \mathcal{J}_{S,\Lambda}(k(K_X + \Delta))) \to H^0(S, \mathcal{J}_{\Lambda|_S}(k(K_X + \Delta))).$$

Since (X, Δ) is plt, (S, Ω) klt and so $(S, \Omega + \frac{k-1}{m}H|_S)$ is klt as m is sufficiently big. Therefore $\mathcal{J}_{\Omega + \frac{k-1}{m}H|_S} = \mathcal{O}_S$. Since

$$\Lambda|_S - (\Sigma + k\Phi) = \left(\frac{k-1}{m}G + B\right)\Big|_S - (\Sigma + k\Phi) \leq \Omega + \frac{k-1}{m}H|_S,$$

then by (2) of (6.4), we have $\mathscr{I}_{\Sigma+k\Phi} \subset \mathscr{J}_{\Lambda|_S}$ and so

$$\Sigma + k\Phi \in |k(K_X + \Delta)|_S$$

and the theorem follows.

Let $f: Y \to X$ be a log resolution of $(X, \Delta + \frac{1}{k}A)$ and of $|l\left(K_X + \Delta + \frac{1}{k}A\right)|$. We write

$$K_Y + \Gamma = f^*\left(K_X + \Delta + \frac{1}{k}A\right) + E$$

as in (3.21.1). Define

$$\Xi = \Gamma - \Gamma \wedge \frac{\operatorname{Fix}(l(K_Y + \Gamma))}{l}.$$

We have that $l(K_Y + \Xi)$ is Cartier, $\operatorname{Fix}(l(K_Y + \Xi)) \wedge \Xi = 0$ and $\operatorname{Mob}(l(K_Y + \Xi))$ is free. Since $\operatorname{Fix}(l(K_Y + \Xi)) + \Xi$ has simple normal crossings support, it follows that $\mathbf{B}(K_Y + \Xi)$ contains no non-klt center of $(Y, \lceil \Xi \rceil)$. Let $T = f_*^{-1}S$. Let $\Gamma_T = (\Gamma - T)|_T$ and $\Xi_T = (\Xi - T)|_T$. Let $m > 0$ be divisible by l. By (6.20)

$$H^0(T, \mathscr{O}_T(m(K_T + \Xi_T))) = H^0(T, \mathscr{J}_{\|m(K_T+\Xi_T)\|}(m(K_T + \Xi_T))).$$

By (6.11), there is an ample divisor H on Y (independent of m) such that if $\tau \in H^0(T, \mathscr{O}_T(H))$ and $\sigma \in H^0(T, \mathscr{O}_T(m(K_T + \Xi_T)))$, then $\sigma \cdot \tau$ is in the image of the homomorphism

$$H^0(Y, \mathscr{O}_Y(m(K_Y + \Xi) + H)) \to H^0(T, \mathscr{O}_T(m(K_Y + \Xi) + H)).$$

Therefore

$$(\ddagger) \qquad |m(K_T + \Xi_T)| + m(\Gamma_T - \Xi_T) + |H|_T| \subset |m(K_Y + \Gamma) + H|_T.$$

We claim that

$$(\flat) \qquad \Omega + \frac{1}{k}A|_S \geq (f|_T)_*\Xi_T \geq \Omega - \Phi + \frac{1}{k}A|_S$$

and so, as $(S, \Omega + \frac{1}{k}A|_S)$ is canonical, we have

$$\begin{aligned} &|m(K_S + \Omega - \Phi)| + m((f|_T)_*\Xi_T - \Omega + \Phi) \\ &\qquad \subset |m(K_S + (f|_T)_*\Xi_T)| = (f|_T)_*|m(K_T + \Xi_T)|, \end{aligned}$$

cf. (6.28). Pushing forward the inclusion ($\ddagger$), one sees that

$$|m(K_S + \Omega - \Phi)| + m\Phi + \left(\frac{m}{k}A + f_*H\right)\Big|_S \subset |m(K_X + \Delta) + \frac{m}{k}A + f_*H|_S.$$

Equation (†) now follows.

We will now prove the inequality ($\flat$) claimed above. We have $\Xi_T \leq \Gamma_T$ and $(f|_T)_*\Gamma_T = \Omega + \frac{1}{k}A|_S$ and so the first inequality follows.

In order to prove the second inequality, let P be any prime divisor on S and let P' be its strict transform on T. Assume that P is contained in the support of Ω, or equivalently, that P' is contained in the support of $(\Gamma - T)|_T$. Then there is a component G of the support of Γ such that

$$\mathrm{mult}_{P'}(\mathrm{Fix}|l(K_Y + \Gamma)|_T) = \mathrm{mult}_G(\mathrm{Fix}|l(K_Y + \Gamma)|)$$

and $\mathrm{mult}_{P'}(\Gamma_T) = \mathrm{mult}_G(\Gamma)$. It follows that

$$\mathrm{mult}_{P'}(\Xi_T) = \mathrm{mult}_G(\Xi) = \mathrm{mult}_G(\Gamma) - \min\left\{\mathrm{mult}_G(\Gamma),\ \frac{1}{l}\mathrm{mult}_G(\mathrm{Fix}|l(K_Y + \Gamma)|)\right\}.$$

Since $\mathrm{mult}_P(\Omega + \frac{1}{k}A|_S) = \mathrm{mult}_{P'}(\Gamma_T)$ and $\Omega \wedge (1 - \frac{\varepsilon}{k})F \le \Phi$, it suffices to show that

$$\mathrm{mult}_{P'}(\mathrm{Fix}\,|l(K_Y + \Gamma)|_T) \le l\left(1 - \frac{\varepsilon}{k}\right)\mathrm{mult}_P(F).$$

Since $E|_T$ is exceptional, we have that

$$\mathrm{mult}_{P'}(\mathrm{Fix}\,|l(K_Y + \Gamma)|_T) = \mathrm{mult}_P\left(\mathrm{Fix}\left|l\left(K_X + \Delta + \frac{1}{k}A\right)\right|_S\right)$$

cf. (6.29). Let $\eta > \varepsilon/k$ be a rational number such that $\eta(K_X + \Delta) + \frac{1}{k}A$ is ample. Since l is sufficiently big and divisible and

$$l(K_X + \Delta + \frac{1}{k}A) \simeq l(1-\eta)(K_X + \Delta) + l(\eta(K_X + \Delta) + \frac{1}{k}A),$$

we have that

$$\mathrm{mult}_P\left(\mathrm{Fix}\left|l\left(K_X + \Delta + \frac{1}{k}A\right)\right|_S\right) \le l\left(1 - \frac{\varepsilon}{k}\right)\mathrm{mult}_P(F)$$

where we have used the fact that if $\mathrm{mult}_P(F) = 0$, then $\mathrm{mult}_P(\mathrm{Fix}\,|l(K_X + \Delta + \frac{1}{k}A)|_S) = 0$ for all $l \gg 0$ cf. (6.21). □

Corollary 6.14. *Let $\pi\colon X \to Z$ be a projective morphism from a smooth quasi-projective variety to an affine variety. Let $\Delta = S + A + B$ be a $\mathbb{Q}$-divisor such that $S = \lfloor\Delta\rfloor$ is irreducible and smooth, $A \ge 0$ is a general ample $\mathbb{Q}$-divisor, $B \ge 0$, (X, Δ) is plt, (S, Ω) is terminal where $\Omega = (A + B)|_S$, and $\mathbf{B}(K_X + \Delta)$ does not contain S. For any sufficiently divisible integer $m > 0$ let*

$$F_m = \mathrm{Fix}(|m(K_X + \Delta)|_S)/m \qquad \text{and} \qquad F = \lim F_{m!}.$$

If Φ is a $\mathbb{Q}$-divisor on S and $k > 0$ is an integer such that $k\Delta$ and $k\Phi$ are Cartier and

$$\Omega \wedge F \le \Phi \le \Omega,$$

then

$$|k(K_S + \Omega - \Phi)| + k\Phi \subset |k(K_X + \Delta)|_S.$$

Proof. Since (S,Ω) is terminal, $(S,\Omega+A')$ is also terminal for $A'=\delta A$ where $0<\delta\ll 1$. Fix $1\gg\varepsilon>0$ such that $\varepsilon(K_X+\Delta)+A'$ is ample. Clearly $\Omega\wedge(1-\varepsilon/k)F\le\Phi\le\Omega$ and the result follows from (6.12). □

REMARK 6.15. The following example shows that in some cases

$$|k(K_S+\Omega)|\neq|k(K_X+\Delta)|$$

and hence it is indeed necessary to replace Ω by $\Omega-\Omega\wedge F$.

Let X be the blow-up of $\mathbb{P}^2$ at a point q, E the exceptional divisor, S the strict transform of a line through q and L_i the strict transforms of general lines on $\mathbb{P}^2$. We let $\Delta=S+\frac{2}{3}(L_1+L_2+L_3)$ so that $\Omega=\frac{2}{3}(E+L_1+L_2+L_3)|_S=\frac{2}{3}(p+p_1+p_2+p_3)$. For $k=3l$, we have that $|k(K_S+\Omega)|=|2lp|=\mathbb{P}^{2l}$ whereas $|k(K_X+\Delta)|=|2lE|$.

REMARK 6.16. The topics treated in this section can be found in [HM06], [HM07], [HM08]. They are based on ideas of [Siu98], [Siu04], [Kaw99b], [Kaw99a], [Tsu07], [Tsu04]. Related results may be found in [Amb06], [MV07], [Var08], [Tak06], [Tak07], [Pău07], [BP08], [Laz09].

EXERCISE 6.17. Notation as in (6.4). Show that if $c<1$ and $D\in V$ is general, then $\mathcal{J}_{\Delta,c\cdot V}=\mathcal{J}_{\Delta,cD}$.

EXERCISE 6.18. Notation as in (6.4). Show that $\mathcal{J}_{\Delta,D}=\mathcal{O}_X$ if and only if $(X,D+\Delta)$ is dlt and $\lfloor D\rfloor=0$.

EXERCISE 6.19. Notation as in (6.4). Show that $\mathcal{J}_{\Delta,cD}=\mathcal{O}_X$ if $c\ll 1$ (in fact $\mathcal{J}_{cD}=\mathcal{O}_X$ if $\operatorname{mult}_x(cD)<1$ for all $x\in X$ cf. [Laz04b, 9.5.13]) and that $\mathcal{J}_{\Delta,D}\subset m_x$ if $\operatorname{mult}_x(D)\ge\dim X$.

EXERCISE 6.20. Let $\pi\colon X\to Z$ be a projective morphism from a smooth quasi-projective variety to an affine variety. Let $D\ge 0$ be a divisor on X. Show that $H^0(\mathcal{O}_X(D))=H^0(\mathcal{J}_{\|D\|}(D))$.

EXERCISE 6.21. Let $D\ge 0$ be a Cartier divisor on a smooth variety X and $Z\subset X$ a closed subvariety such that $\lim\operatorname{mult}_Z(|m!D|)/m!=0$. Show that $Z\not\subset\mathbf{B}_-(D)$.

EXERCISE 6.22. Use the arguments of [Laz04b, 9.2.19] to give an alternative proof of the claim in (6.3).

EXERCISE 6.23. Show that $\mathcal{J}_{\Delta,D}=\mathcal{J}_{\Delta,\{D\}}\otimes\mathcal{O}_X(-\lfloor D\rfloor)$.

EXERCISE 6.24. Let $f\colon X\to Y$ be a birational morphism of normal varieties. If Y is $\mathbb{Q}$-factorial, show that there exists an effective f-exceptional divisor F such that $-F$ is ample over Y.

EXERCISE 6.25. Let $f\colon Y\to X$ and $\pi\colon X\to Z$ be projective morphisms of normal quasi-projective varieties. Show that if $\mathcal{F}$ is a coherent sheaf on Y such that $\mathcal{R}^j f_*\mathcal{F}=0$ for all $j>0$, then $\mathcal{R}^i(\pi\circ f)_*\mathcal{F}\simeq\mathcal{R}^i\pi_*(f_*\mathcal{F})$ for all $i>0$.

EXERCISE 6.26. Let $\pi\colon X\to Z$ be a projective morphism from a smooth variety to an affine variety. Let $\mathcal{F}$ be a coherent sheaf on X and H be a very ample divisor on X such that $H^i(X,\mathcal{F}(mH))=0$ for all $i>0$ and all $m\ge-\dim X$. Show that $\mathcal{F}$ is generated by global sections.

EXERCISE 6.27. Show that if $\Delta = \sum \delta_j \Delta_j$, then the divisors Δ^i in (6.11) are given by $\Delta^i = \sum_{j|\delta_j > (k-s)/k} \Delta_j$.

EXERCISE 6.28. Let $f: Y \to X$ be a proper birational morphism of normal varieties. Show that if $\Xi \geq 0$ is a $\mathbb{Q}$-divisor on Y such that $(X, f_*\Xi)$ is canonical and $m > 0$ is an integer such that $m(K_X + f_*\Xi)$ and $m(K_Y + \Xi)$ are Cartier, then $|m(K_X + f_*\Xi)| = f_*|m(K_Y + \Xi)|$.

EXERCISE 6.29. Let $f: Y \to X$ be a proper birational morphism of normal varieties, $S \subset X$ a divisor and $T = f_*^{-1}S$. Let D be a Cartier divisor on X. Show that $|f^*D|_T = (f|_T)^*|D|_S$. Assume that T and S are normal and that P is a prime divisor on S and P' is its strict transform on T. Show that $\operatorname{mult}_{P'}(\operatorname{Fix}|f^*D|_T) = \operatorname{mult}_P(\operatorname{Fix}|D|_S)$.

CHAPTER 7

Finite generation of the restricted algebra

7.A Rationality of the restricted algebra

Theorem 7.1. *Assume Theorem 5.60 in dimension $\leq n-1$. Let $\pi\colon X \to Z$ be a projective morphism from a smooth quasi-projective variety of dimension n to a normal affine variety. Let $\Delta = S + A + B$ be a $\mathbb{Q}$-divisor such that $S = \lfloor \Delta \rfloor$ is irreducible and smooth, $A \geq 0$ is a general ample $\mathbb{Q}$-divisor, $B \geq 0$, (X, Δ) is plt, $(S, \Omega + A|_S)$ is canonical where $\Omega = (A+B)|_S$, and $\mathbf{B}(K_X + \Delta)$ does not contain S. For any sufficiently divisible integer $m > 0$ let*

$$F_m = \operatorname{Fix}(|m(K_X + \Delta)|_S)/m \qquad \text{and} \qquad F = \lim F_{m!}.$$

Then $\Theta = \Omega - \Omega \wedge F$ is rational and if $k\Delta$ and $k\Theta$ are Cartier, then

$$R_S(k(K_X + \Delta)) \simeq R(k(K_S + \Theta)).$$

Proof. Let $V \subset \operatorname{WDiv}_{\mathbb{R}}(S)$ be the sub-vector space generated by the components of Ω and $W \subset V$ be the smallest rational affine subspace containing Θ. By (5.60), there exists a constant $\delta > 0$ such that if $\Theta' \in V$ with $\|\Theta' - \Theta\| \leq \delta$, then any prime divisor contained in $\mathbf{B}(K_S + \Theta)$ is also contained in $\mathbf{B}(K_S + \Theta')$ and there exist a constant $\eta > 0$ and an integer $r > 0$ such that if $\Theta' \in W$ with $\|\Theta' - \Theta\| \leq \eta$ and $k > 0$ is an integer such that $k(K_S + \Theta')/r$ is Cartier, then every component of $\operatorname{Fix}(k(K_S + \Theta'))$ is contained in $\mathbf{B}(K_S + \Theta)$.

Note that if $l(K_X + \Delta)$ is Cartier and $\Theta_l = \Omega - \Omega \wedge F_l$, then $l(K_S + \Theta_l)$ is Cartier and

$$|l(K_X + \Delta)|_S \subset |l(K_S + \Theta_l)| + l(\Omega \wedge F_l).$$

It follows that no component of $\mathrm{Supp}(\Theta_l)$ is contained in $\mathrm{Fix}(l(K_S + \Theta_l))$ and hence in $\mathbf{B}(K_S + \Theta_l)$. For $l \gg 0$, we have that $\Theta_l \in V$, $\|\Theta_l - \Theta\| \leq \delta$ so that no component of $\mathrm{Supp}(\Theta_l)$ is contained in $\mathbf{B}(K_S + \Theta)$. Since $\mathrm{Supp}(\Theta) \subset \mathrm{Supp}(\Theta_l)$, no component of Θ is contained in $\mathbf{B}(K_S + \Theta)$.

Let $0 < \varepsilon \ll 1$ be a rational number such that $\varepsilon(K_X + \Delta) + A$ is ample. If Θ is not rational, then there exists a component P of Θ with $\mathrm{mult}_P(\Theta) \notin \mathbb{Q}$. Recall that $\Theta = \Omega - \Omega \wedge F$. Notice that if Q is a prime divisor on S such that $\mathrm{mult}_Q(\Omega \wedge F) \notin \mathbb{Q}$, then

$$\mathrm{mult}_Q(\Omega \wedge F) = \mathrm{mult}_Q(F)$$

and if $\mathrm{mult}_Q(\Omega \wedge F) \in \mathbb{Q}$, then $\mathrm{mult}_Q(\Phi) = \mathrm{mult}_Q(\Theta)$ for any $\mathbb{R}$-divisor $\Phi \in W$.

Let W' be the smallest rational affine subspace of $\mathrm{WDiv}_{\mathbb{R}}(S)$ containing $\Omega - \Theta$ so that $W' = \Omega - W$. By Diophantine approximation (see (7.2) below), there exist an integer $k > 0$, an effective $\mathbb{Q}$-divisor $\Phi \in W'$ such that $k\Phi/r$ and $k\Delta/r$ are Cartier, $\mathrm{mult}_P(\Phi) < \mathrm{mult}_P(\Omega \wedge F)$ and $\|\Phi - (\Omega \wedge F)\| \leq \min\{\eta, \gamma, \varepsilon f/k\}$ where f is the smallest non-zero coefficient of F and γ is the smallest non-zero coefficient of $\Omega - \Omega \wedge F$. Then

$$\Omega \wedge (1 - \frac{\varepsilon}{k})F \leq \Phi \leq \Omega.$$

By (6.12), we have that

$$|k(K_S + \Omega - \Phi)| + k\Phi \subset |k(K_X + \Delta)|_S. \tag{b}$$

Since

$$\mathrm{mult}_P(\mathrm{Fix}(|k(K_X + \Delta)|_S)) \geq \mathrm{mult}_P(kF) = \mathrm{mult}_P(k(\Omega \wedge F)) > \mathrm{mult}_P(k\Phi),$$

it follows from (b) that P is contained in the support of $\mathrm{Fix}(k(K_S + \Omega - \Phi))$. Since $||(\Omega - \Phi) - \Theta|| = ||\Phi - (\Omega \wedge F)|| \leq \eta$ and $k(K_S + \Omega - \Phi)/r$ is Cartier, P is contained in $\mathbf{B}(K_S + \Theta)$. This is a contradiction and therefore Θ is rational. The isomorphism $R_S(k(K_X + \Delta)) \simeq R(k(K_S + \Theta))$ now follows from (6.12). □

Lemma 7.2. *Let V be an $\mathbb{R}$-vector space defined over $\mathbb{Q}$. Fix $k \in \mathbb{Z}_{>0}$ and $\alpha \in \mathbb{R}_{>0}$. If $v \in V$, then we may find vectors $v_1, \dots, v_r \in V$ and integers $m_1, \dots, m_r \in \mathbb{Z} > 0$ such that*

(7.2.1) $v = \sum r_i v_i$ where $r_i \in \mathbb{R}_{\geq 0}$ and $\sum r_i = 1$;
(7.2.2) $\|v_i - v\| < \alpha/m_i$; and
(7.2.3) $m_i v_i/k$ is integral.

Proof. See §3 of [BCHM10]. □

7.B Proof of (5.69)

Proof of (5.69). We may assume that Z is affine cf. (5.22). By (5.67), it suffices to show that the restricted algebra $R_S(K_X + \Delta)$ is finitely generated. By (5.68), this is equivalent to showing that the restricted algebra $R_S(k(K_X + \Delta))$ is finitely generated for some integer $k > 0$. It follows that we may replace Δ by a $\mathbb{Q}$-linearly equivalent divisor Δ'. By (5.72), S is mobile. Since f is birational and Z is affine, $\Delta - S$ is big so that $\Delta - S \sim_{\mathbb{Q}} A' + B'$ where A' is ample, $B' \geq 0$ and $\mathrm{Supp}(B')$ does not contain S. Let $\varepsilon > 0$ be a sufficiently small rational number. We may replace Δ by

$$\Delta' = S + A + B \sim_{\mathbb{Q}} \Delta$$

where $A \sim_{\mathbb{Q}} \varepsilon A'$ is a general ample $\mathbb{Q}$-divisor and $B = (1 - \varepsilon)(\Delta - S) + \varepsilon B'$. Let $f: Y \to X$ be a log resolution of (X, Δ') and write

$$K_Y + \Gamma' = f^*(K_X + \Delta') + E'$$

as in (3.21.1). Let $T = f_*^{-1}S$. We may assume that (Y, Γ') is plt and that $(T, (\Gamma' - T)|_T)$ is terminal cf. (3.36). Let $F \geq 0$ be an effective exceptional $\mathbb{Q}$-divisor such that $f^*A - F$ is ample, $(Y, \Gamma' + F)$ is plt and $(T, (\Gamma' - T + F)|_T)$ is terminal. We set $\Gamma = \Gamma' - f^*A + F + H$ where $H \sim_{\mathbb{Q}} f^*A - F$ is a general ample $\mathbb{Q}$-divisor. Then (Y, Γ) is plt and we may assume that $(T, (\Gamma - T + H)|_T)$ is terminal. Since $\mathbf{B}(K_X + \Delta)$ does not contain S, one sees that $\mathbf{B}(K_Y + \Gamma)$ does not contain T. By (7.1), it then follows that there is a $\mathbb{Q}$-divisor Θ on T such that $0 \leq \Theta \leq (\Gamma - T)|_T$ and $R_T(k(K_Y + \Gamma)) \simeq R(k(K_T + \Theta))$ for some $k > 0$ sufficiently divisible. By (5.60), $R(k(K_T + \Theta))$ is finitely generated. The theorem now follows since for $k > 0$ sufficiently divisible,

$$R_T(k(K_Y + \Gamma)) \simeq R_S(k(K_X + \Delta)). \qquad \square$$

REMARK 7.3. Many of the ideas of this section were inspired by [Sho03]. For related results, see [Amb06], [Cor07], [Bir03], [CKL08].

CHAPTER 8

Log terminal models

In this chapter we will prove the following result:

Theorem 8.1. *Theorems 5.56, 5.57, 5.58, 5.59, 5.60 in dimensions $\leq n-1$ and Theorem 5.56 in dimensions $\leq n$ imply Theorem 5.57 in dimension n.*

Proof. See (8.B). □

8.A Special termination

Lemma 8.2. *Assume Theorems 5.56, 5.57, 5.58, 5.59, 5.60 in dimensions $\leq n-1$. Let $\pi : X \to U$ be a projective morphism of normal quasi-projective varieties. Let $(X, \Delta + C = S + A + B + C)$ be a $\mathbb{Q}$-factorial dlt pair with $S = \lfloor \Delta \rfloor$, such that $K_X + \Delta + C$ is nef over U and $\mathbf{B}_+(A/U)$ contains no non-klt centers of $(X, \Delta + C)$. Let $\phi_i : X_i \dashrightarrow X_{i+1}$ be a sequence of flips and divisorial contractions over U for the $(K_X + \Delta)$-mmp over U with scaling of C.*

Then there exists an integer $k > 0$ such that for all $i \geq k$, ϕ_i is an isomorphism on a neighborhood of S.

Proof. Suppose that this is not the case, then we may assume that infinitely often ϕ_i is a flip and the flipping locus intersects a component S' of S.

Step 1. *We may assume that S is irreducible and A is a general ample $\mathbb{Q}$-divisor over U.*

Since $\mathbf{B}_+(A/U)$ contains no non-klt centers of $(X, \Delta + C)$, we have $A \sim_{\mathbb{R},U} A' + B'$ where A' is a $\mathbb{Q}$-divisor ample over U and $\operatorname{Supp}(B')$ contains no non-klt centers of $(X, \Delta + C)$. For any rational number $0 < \varepsilon \ll 1$, $A' + \varepsilon(S - S')$ is an ample $\mathbb{Q}$-divisor.

Let $A'' \sim_{\mathbb{R},U} A' + \varepsilon(S - S')$ be a general ample $\mathbb{Q}$-divisor over U. For any rational number $0 < \delta \ll 1$, we have

$$S + A + B \sim_{\mathbb{R},U} S' + \delta A'' + (1 - \varepsilon\delta)(S - S') + (1 - \delta)A + \delta B' + B$$

where $\delta A''$ is a general ample $\mathbb{Q}$-divisor over U and

$$(X, S' + \delta A'' + (1 - \varepsilon\delta)(S - S') + (1 - \delta)A + \delta B' + B + C)$$

is dlt. Therefore, we may assume that S is irreducible and A is a general ample $\mathbb{Q}$-divisor over U.

Let S_i, A_i, B_i, Δ_i and C_i be the strict transforms of S, A, B, Δ and C on X_i. Note that $(X_i, S_i + A_i + B_i)$ is plt and $\lfloor S_i + A_i + B_i \rfloor = S_i$. Therefore, S_i is normal, cf. (3.24). We write $(K_{X_i} + \Delta_i)|_{S_i} = K_{S_i} + \Delta_{S_i}$. Note that (S_i, Δ_{S_i}) is klt.

Step 2. *We may assume that for $i \gg 0$, the induced birational map $\psi_i : S_i \dashrightarrow S_{i+1}$ is an isomorphism in codimension 1 and $(\psi_i)_*\Delta_{S_i} = \Delta_{S_{i+1}}$.*

We follow closely section 4.2 of [Fuj07]. Let $\Delta = \sum_{i \in I} \delta_i \Delta_i$, $d =$ totaldiscrep$(S, \Delta_S) \in (-1, 0]$, and

$$\mathcal{B} = \left\{ 1 - \frac{1}{m} + \sum_{i \in I} \frac{r_i \delta_i}{m} \in [0, -d] | m \in \mathbb{Z}_{>0},\ r_i \in \mathbb{Z}_{\geq 0} \right\}.$$

Then $\mathcal{B}$ is a finite set and there is an integer $0 \leq d_{\mathcal{B}} < \infty$ defined by

$$d_{\mathcal{B}}(S, \Delta_S) := \sum_{\beta \in \mathcal{B}} \sharp\{E | a(E; S, \Delta_S) < -\beta\}$$

cf. [Fuj07]. Note that by adjunction, we have that $a(E; S_i, \Delta_{S_i}) \leq a(E; S_{i+1}, \Delta_{S_{i+1}})$ for any divisor E over S. Therefore $d_{\mathcal{B}}(S_i, \Delta_{S_i}) \geq d_{\mathcal{B}}(S_{i+1}, \Delta_{S_{i+1}})$ and so we may assume that $d_{\mathcal{B}}(S_i, \Delta_{S_i})$ is constant for $i \gg 0$.

Suppose that $P \subset S_{i+1}$ is a divisor such that $(\psi_i^{-1})_* P = 0$. The above inequality is strict because $-a(P; S_{i+1}, \Delta_{S_{i+1}}) \in \mathcal{B}$ cf. (3.24). Therefore $S_i \dashrightarrow S_{i+1}$ extracts no divisors for all $i \gg 0$. If $S_i \dashrightarrow S_{i+1}$ contracts a divisor, then $\varrho(S_i) > \varrho(S_{i+1})$. This can happen at most finitely many times and so $S_i \dashrightarrow S_{i+1}$ is an isomorphism in codimension 1 for all $i \gg 0$. Notice that if $(\psi_i)_*\Delta_{S_i} \neq \Delta_{S_{i+1}}$, then $(\psi_i)_*\Delta_{S_i} \geq \Delta_{S_{i+1}}$ cf. (8.4) and $d_{\mathcal{B}}(S_i, \Delta_{S_i}) > d_{\mathcal{B}}(S_{i+1}, \Delta_{S_{i+1}})$. This is a contradiction and hence Step 2 follows.

Step 3. *There exist integers $j > i$ such that $S_i \dashrightarrow S_{i+1}$ is not an isomorphism but $S_i \dashrightarrow S_j$ is an isomorphism.*

By assumption, there is a non-increasing sequence of numbers $\lambda_i \geq 0$ such that $K_X + \Delta_i + \lambda_i C_i$ is nef over U. Since $K_{S_i} + \Delta_{S_i} + \lambda_i C_i|_{S_i}$ is nef over U and $S_i \dashrightarrow S_{i+1}$ is an isomorphism in codimension 1, by (5.76), there are finitely many rational maps $S_i \dashrightarrow S_j$ (modulo isomorphism). Let $X_i \to Z_i$ be the flipping contraction. Let T_i be the normalization of the image of S_i in Z_i. Suppose that $S_i \to T_i$ is an isomorphism and the flipping locus intersects S_i; then $\Sigma_i \cdot S_i > 0$ for any flipping curve $\Sigma_i \subset X_i$ and so $\Sigma_{i+1} \cdot S_{i+1} < 0$

for any flipped curve $\Sigma_{i+1} \subset X_{i+1}$ and so $S_{i+1} \to T_i$ is not an isomorphism. Therefore, we may assume that infinitely often $S_i \to T_i$ or $S_{i+1} \to T_i$ is not an isomorphism. It follows that there is a divisor F over S_i such that $a(F; S_i, \Delta_{S_i}) < a(F; S_{i+1}, \Delta_{S_{i+1}})$ cf. (8.4). Therefore, infinitely often $S_i \dashrightarrow S_{i+1}$ is not an isomorphism. Step 3 now follows.

Step 4. *We find the required contradiction.*

Let F be a divisor over S_i with center contained in the flipping or flipped locus of $X_i \dashrightarrow X_{i+1}$, then as we have seen above $a(F; S_i, \Delta_{S_i}) < a(F; S_{i+1}, \Delta_{S_{i+1}})$. Since $a(F; S_{i+1}, \Delta_{S_{i+1}}) \le a(F; S_j, \Delta_{S_j})$, we have the required contradiction. □

Lemma 8.3. *Assume Theorems 5.56, 5.57, 5.58, 5.59, 5.60 in dimensions $\le n-1$ and Theorem 5.56 in dimensions $\le n$. Let $\pi : X \to U$ be a projective morphism of normal quasi-projective varieties. Let $(X, \Delta + C = S + A + B + C)$ be a $\mathbb{Q}$-factorial dlt pair with $S = \lfloor \Delta \rfloor$, such that $K_X + \Delta + C$ is nef over U and $\mathbf{B}_+(A/U)$ contains no non-klt centers of $(X, \Delta + C)$. Suppose that there is an effective $\mathbb{R}$-divisor D such that* Supp$(D) \subset S$ *and a non-negative number α such that*

$$(\sharp) \qquad K_X + \Delta \sim_{\mathbb{R},U} D + \alpha C.$$

Then there exists an ltm $\phi: X \dashrightarrow Y$ for $K_X + \Delta$ over U such that $\mathbf{B}_+(\phi_* A/U)$ contains no non-klt centers of $(Y, \phi_*\Delta)$.

Proof. We run the $(K_X + \Delta)$-mmp over U with scaling of C. If $R = \mathbb{R}^+[\Sigma]$ is a $(K_X + \Delta)$-negative extremal ray such that $(K_X + \Delta) \cdot \Sigma < 0$ and $C \cdot \Sigma > 0$, then $D \cdot \Sigma < 0$ and so R induces a flipping or divisorial contraction.

Notice that if $f_i : X_i \to Z_i$ is a flipping contraction and $R_i = \mathbb{R}^+[\Sigma_i]$ is the corresponding negative extremal ray, as $D_i \cdot \Sigma_i < 0$, there is a component S_i' of S_i such that $S_i' \cdot \Sigma_i < 0$. But then f_i is a pl-flipping contraction for some dlt pair $K_{X_i} + \Delta_i'$ where $\lfloor \Delta_i' \rfloor = S_i'$ cf. (5.71). Therefore, by Theorem 5.56 in dimension n, the flip $X_i \dashrightarrow X_{i+1}$ exists.

Note that condition $(\sharp)$ is preserved by flips and divisorial contractions. Since $S_i' \cdot \Sigma_i < 0$, $X_i \to Z_i$ is not an isomorphism on a neighborhood of S_i. By (8.2), the sequence of flips and divisorial contractions terminates. □

8.B Existence of log terminal models

Proof of (8.1). **Step 1.** *We may assume that X is smooth,* Supp$(\Delta + D)$ *is simple normal crossings, $\Delta = A + B$, where A is a general ample $\mathbb{Q}$-divisor over U and if G is a component of* Supp(D)*, then either G is contained in $\mathbf{B}(D/U)$ or mG is mobile for some integer $m > 0$.*

Replacing Δ by an $\mathbb{R}$-linearly equivalent divisor, we may assume that $\Delta \ge A$ where A is a general ample $\mathbb{Q}$-divisor. Let $f : Y \to X$ be a log resolution of (X, Δ). We write $K_Y + \Gamma = f^*(K_X + \Delta) + E$ as in (3.21.1). We may assume that there exists $F \ge 0$ a $\mathbb{Q}$-divisor such that Supp(F) is the exceptional set and $K_Y + \Gamma + F$ is klt. By (5.42), an ltm over U for $K_Y+\Gamma+F$ is an ltm over U for $K_X+\Delta$. There exists an exceptional $\mathbb{Q}$-divisor $\Phi\geq 0$ such that $f^*A-\Phi$ is ample and $(Y,\Gamma+F+\Phi)$ is klt. Let $C\sim_{\mathbb{Q},U} f^*A-\Phi$ be a general ample $\mathbb{Q}$-divisor over U. Since $\Gamma+F+\Phi-f^*A+C\sim_{\mathbb{Q},U}\Gamma+F$, we may replace $(Y,\Gamma+F)$ by $(Y,\Gamma'=\Gamma+F+\Phi-f^*A+C)$. Note that $K_Y+\Gamma'\sim_{\mathbb{R},U} D'=f^*D+F+E\geq 0$. By (2.9), we may assume that all components G of $\mathrm{Supp}(D')$ are either contained in $\mathbf{B}(D'/U)$ or mG is mobile for some $m>0$.

Step 2. *There exists a rational map $\phi\colon X \dashrightarrow Y$ over U such that*

(8.3.1) ϕ extracts no divisors;

(8.3.2) if P is a prime divisor contracted by ϕ, then $P\subset \mathrm{Supp}(D)$;

(8.3.3) Y is $\mathbb{Q}$-factorial and $Y\to U$ is projective;

(8.3.4) $K_Y+\phi_*\Delta$ is dlt and nef over U; and

(8.3.5) no non-klt center of $K_Y+\phi_*\Delta$ is contained in $\mathbf{B}_+(\phi_*A/U)$.

We write $D=D_1+D_2$ where the components of D_1 are contained in $\mathrm{Supp}(\lfloor\Delta\rfloor)$ and the components of D_2 are not contained in $\mathrm{Supp}(\lfloor\Delta\rfloor)$. We proceed by induction on the number of components of D_2.

If $D_2=0$, then pick H a general element of a sufficiently ample divisor. We may assume that $K_X+\Delta+H$ is dlt and nef over U. We may then run the $(K_X+\Delta)$-mmp over U with scaling of H. By (8.3), there exists an ltm over U, $\phi\colon X\dashrightarrow Y$. It is easy to see that ϕ satisfies properties (1-5) above.

If $D_2\neq 0$, let

$$\lambda=\sup\{t\geq 0\,|\,(X,\Delta+tD_2)\text{ is dlt}\}.$$

Then $(X,\Delta'=\Delta+\lambda D_2)$ is dlt and

$$K_X+\Delta'\sim_{\mathbb{R},U} D'=D+\lambda D_2.$$

Note that $\mathrm{Supp}(D')=\mathrm{Supp}(D)$ and the number of components of D' not contained in $\lfloor\Delta'\rfloor$ is less than the number of components of D not contained in $\lfloor\Delta\rfloor$. By induction, there is a rational map $\phi'\colon X\dashrightarrow Y'$ over U such that Δ' and D' satisfy properties (1-5) above. Since $\lambda>0$, since $K_{Y'}+\phi'_*(\Delta+\lambda D_2)$ is dlt and nef over U, since $\mathbf{B}_+(\phi'_*A/U)$ contains no non-klt center of $(Y',\phi'_*(\Delta+\lambda D_2))$ and since

$$K_{Y'}+\phi'_*\Delta\sim_{\mathbb{R},U}\phi'_*D_1+\phi'_*D_2,$$

then by (8.3), there exists an ltm $f\colon Y'\dashrightarrow Y$ for $K_{Y'}+\phi'_*\Delta$ over U. Let $\phi=f\circ\phi'\colon X\dashrightarrow Y$. Then ϕ satisfies properties (1-5) above.

Step 3. *$\phi\colon X\dashrightarrow Y$ is an ltm for (X,Δ) over U.*

Note that by Step 2, ϕ only contracts components of D, but as ϕ can not contract movable divisors, by Step 1, ϕ only contracts divisors in $\mathbf{B}(K_X+\Delta/U)$.

Let $p\colon W\to X$ and $q\colon W\to Y$ be proper birational maps of normal varieties such that $\phi=q\circ p^{-1}$. Then

$$p^*(K_X+\Delta)+E=q^*(K_Y+\phi_*\Delta)+F$$

where E and F are effective q-exceptional $\mathbb{R}$-divisors with no common components.

We have that $K_Y + \phi_*\Delta$ is semiample over U cf. (5.10). Therefore,

$$\mathbf{B}(p^*(K_X + \Delta) + E/U) = \mathbf{B}(q^*(K_Y + \phi_*\Delta) + F/U) = \mathrm{Supp}(F).$$

On the other hand p_*E is ϕ-exceptional and thus it is contained in $\mathbf{B}(K_X + \Delta/U)$ and therefore

$$\mathrm{Supp}(E) \subset \mathbf{B}(p^*(K_X + \Delta) + E - p_*^{-1}p_*E/U) \subset \mathbf{B}(p^*(K_X + \Delta) + E/U)$$

(where the last inclusion follows by (8.5)). Therefore, as E and F have no common components, we deduce that $E = 0$ and $p_*(F)$ contains all ϕ-exceptional divisors. It follows that $a(P; X, \Delta) < a(P; Y, \phi_*\Delta)$ for all ϕ exceptional divisors $P \subset X$ and so Y is an ltm for (X, Δ) over U. □

EXERCISE 8.4. Let (X, Δ) and (X', Δ') be plt pairs such that $S = \lfloor\Delta\rfloor$ and $S' = \lfloor\Delta'\rfloor$, $\phi : X \dashrightarrow X'$ is a birational map that extracts no divisors, $\Delta' = \phi_*\Delta$, $\phi|_S : S \dashrightarrow S'$ extracts no divisors, and if $p : W \to X$ and $q : W \to X'$ is a common log resolution, then $p^*(K_X + \Delta) = q^*(K_{X'} + \Delta') + F$ where $F \geq 0$. Use (3.24) to show that $(\phi|_S)_*\Delta_S \geq \Delta_{S'}$ and $a(E; S, \Delta_S) \leq a(E; S', \Delta_{S'})$ for any divisor E over S (where $(K_X + \Delta)|_S = K_S + \Delta_S$ and $(K_{X'} + \Delta')|_S = K_{S'} + \Delta_{S'}$).

EXERCISE 8.5. Let $F, D \geq 0$ be effective $\mathbb{R}$-divisors such that $\mathrm{Supp}(F) \subset \mathbf{Fix}(D)$. Show that $\mathbf{Fix}(D + F) = \mathbf{Fix}(D) + F$.

CHAPTER 9

Non-vanishing

In this chapter we will prove that Theorems 5.56, 5.57, 5.58, 5.59, 5.60 in dimensions $\leq n-1$ and Theorems 5.56, 5.57 in dimensions $\leq n$ imply Theorem 5.58 in dimension n. We begin by recalling the following results from [Nak04].

9.A Nakayama-Zariski decomposition

Definition 9.1. Let X be a smooth variety, D a big $\mathbb{R}$-divisor on X and Q a prime divisor on X. Then we define

$$\sigma_Q(D) = \inf\{\operatorname{mult}_Q(D') \mid D' \in |D|_{\mathbb{R}}\}, \qquad \text{and}$$

$$N_\sigma(D) = \sum \sigma_Q(D)Q.$$

REMARK 9.2. By [Nak04, §III.1], it is known that $\sigma_Q(D) = \sigma_Q(D')$ if $D \equiv D'$ and that $\sigma_Q(\ldots)$ is a continuous function on the cone of big divisors.

Definition 9.3. Let X be a smooth variety, D a pseudo-effective $\mathbb{R}$-divisor on X, A an ample $\mathbb{R}$-divisor on X, and Q a prime divisor on X. Then

$$\sigma_Q(D) = \lim_{\varepsilon \to 0^+} \sigma_Q(D + \varepsilon A).$$

REMARK 9.4. By [Nak04, §III.1], $\sigma_Q(D)$ is independent of the choice of A and there are only finitely many prime divisors Q such that $\sigma_Q(D) > 0$. Therefore, $N_\sigma(D) = \sum \sigma_Q(D)Q$ is a well-defined $\mathbb{R}$-divisor. If $D \equiv D'$, then $N_\sigma(D) = N_\sigma(D')$. We have that $D - N_\sigma(D)$ is pseudo-effective and $N_\sigma(D - N_\sigma(D)) = 0$. Finally, if Q is an effective $\mathbb{R}$-divisor whose support is contained in $\operatorname{Supp}(N_\sigma(D))$, then $N_\sigma(D+Q) = N_\sigma(D) + Q$ cf. (8.5).

Lemma 9.5. *Let X be a smooth variety, D a pseudo-effective $\mathbb{R}$-divisor, G a $\mathbb{Q}$-divisor such that $A = G - D$ is an ample $\mathbb{R}$-divisor and Q a prime divisor. If $\sigma_Q(D) = 0$, then Q is not contained in $\mathbf{B}(G)$. In particular Q is not contained in $\mathbf{B}_-(D)$.*

Proof. Let H be a very ample divisor on X. For any $l > 0$ sufficiently divisible we have that $\frac{l}{2}A - (\dim X + 1)H - K_X$ is an ample $\mathbb{R}$-divisor and $l(D + A/2) \sim_{\mathbb{R}} \Delta$ where $\operatorname{mult}_Q(\Delta) < 1$. Therefore $\mathcal{J}_\Delta(lG)$ is generated by global sections cf. (6.26). The lemma now follows as Q is not contained in the co-support of $\mathcal{J}_\Delta$ cf. (6.19). □

One of the key results that we will need is the following.

Theorem 9.6. *Let X be a smooth variety, and D be a pseudo-effective $\mathbb{R}$-divisor on X. Then the following are equivalent:*

(9.6.1) $D \equiv N_\sigma(D)$;

(9.6.2) for any ample divisor A, the function $h^0(X, \mathcal{O}_X(\lfloor tD \rfloor + A))$ is bounded;

(9.6.3) for any ample divisor A, $\lim_{t\to\infty} \frac{1}{t} h^0(X, \mathcal{O}_X(\lfloor tD \rfloor + A)) = 0$.

Proof. We follow [Nak04, V.1.12]. We begin by showing that (1) implies (2). Since $P = N_\sigma(D) - D \equiv 0$, one sees that there exists an ample divisor A' such that $|A' - A + \lfloor tP \rfloor| \neq \emptyset$ for any integer $t \geq 0$. (It suffices in fact to pick A' such that

$$A' - A + \lfloor tP \rfloor - K_X - (n+1)H \equiv A' - A - \{tP\} - K_X - (n+1)H$$

is ample for some very ample divisor H and any integer $t > 0$ and to apply (6.26).) By (2.3) $\lfloor tN_\sigma(D) \rfloor \geq \lfloor tD \rfloor + \lfloor tP \rfloor$, and therefore

$$h^0(\mathcal{O}_X(\lfloor tD \rfloor + A)) \leq h^0(\mathcal{O}_X(\lfloor tN_\sigma(D) \rfloor + A')).$$

For any divisor Q in the support of $N_\sigma(D)$, and any $k \gg 0$, we have that $\sigma_Q(kN_\sigma(D) + A') > 0$. By (9.4), it follows that for $t > k$, we have $\sigma_Q(tN_\sigma(D) + A') > (t-k)\sigma_Q(N_\sigma(D))$. In particular $h^0(\mathcal{O}_X(\lfloor tN_\sigma(D) \rfloor + A')) = h^0(\mathcal{O}_X(\lfloor kN_\sigma(D) \rfloor + A'))$.

Clearly (2) implies (3).

We will now see that (3) implies (1). Since $D - N_\sigma(D)$ is pseudo-effective and

$$h^0(\mathcal{O}_X(\lfloor t(D - N_\sigma(D)) \rfloor + A)) \leq h^0(\mathcal{O}_X(\lfloor tD \rfloor + A))$$

we may replace D by $D - N_\sigma(D)$. We may therefore assume that $N_\sigma(D) = 0$ and we must show that $D \equiv 0$. Let A be a very ample divisor and C be a curve given by the intersection of $n-1 = \dim X - 1$ general elements of $|A|$. We fix an ample divisor A' such that $\frac{1}{2}A' - K_X - \{mD\} - (n-1)A$ is ample for all $m \geq 0$. Let $H_i \in |A|$ be general elements vanishing along C, then $\mathcal{J}(\frac{n-1}{n}(H_1 + \cdots + H_n)) = \mathcal{I}_C$. Since $\mathbf{B}_-(D)$ is the union of countably many subvarieties of codimension at least 2, $\mathbf{B}_-(D) \cap C = \emptyset$. Therefore, there exists an $\mathbb{R}$-divisor $\Delta \sim_{\mathbb{Q}} mD + \frac{1}{2}A'$ such that $\mathcal{J}(\Delta + \frac{n-1}{n}(H_1 + \cdots + H_n)) = \mathcal{I}_C$ on a neighborhood of C cf. (9.22). Therefore, we have that

$$H^1(\mathcal{J}(\Delta + \frac{n-1}{n}(H_1 + \cdots + H_n))(\lfloor mD \rfloor + A')) = 0$$

and so

$$H^0(\mathcal{O}_X(\lfloor mD \rfloor + A')) \to H^0(\mathcal{O}_C(\lfloor mD \rfloor + A'))$$

is surjective. But then $\lim \frac{1}{m} h^0(\mathcal{O}_C(\lfloor mD \rfloor + A')) = 0$ so that $D \cdot C \leq 0$. Since D is pseudo-effective, $D \equiv 0$ cf. (9.23). □

We will also need the following results concerning the Nakayama-Zariski decomposition of the restricted linear series $|D|_S$.

Definition 9.7. Let S be a smooth divisor on a smooth projective variety X, P a prime divisor on S, and D be a big $\mathbb{R}$-divisor on X such that S is not contained in $\mathbf{B}_+(D)$, then we define

$$\sigma_P(\|D\|_S) = \inf\{\operatorname{mult}_P(D'|_S) \big| D \sim_{\mathbb{R}} D' \geq 0,\ S \not\subset \operatorname{Supp}(D')\}, \qquad \text{and}$$

$$N_\sigma(\|D\|_S) = \sum \sigma_P(\|D\|_S) P.$$

We have the following:

Lemma 9.8. *Let S be a smooth divisor on a smooth projective variety X.*

(9.8.1) If A is an ample $\mathbb{R}$-divisor and D is a big $\mathbb{R}$-divisor such that $S \not\subset \mathbf{B}_+(D)$, then $\sigma_P(\|D + A\|_S) \leq \sigma_P(\|D\|_S)$. In particular $N_\sigma(\|A\|_S) = 0$.

(9.8.2) If D is a big $\mathbb{R}$-divisor such that $S \not\subset \mathbf{B}_+(D)$ and P is a prime divisor on S, then $\lim_{\varepsilon \to 0^+} \sigma_P(\|D + \varepsilon A\|_S) = \sigma_P(\|D\|_S)$.

(9.8.3) If D is a pseudo-effective $\mathbb{R}$-divisor such that $S \not\subset \mathbf{B}_-(D)$ and P is a prime divisor on S, then $\lim_{\varepsilon \to 0^+} \sigma_P(\|D + \varepsilon A\|_S) = \sigma_P(\|D\|_S)$ exists and is independent of A.

Proof. The first statement is clear as $A \sim_{\mathbb{R}} \sum r_i A_i$ and $r_i \in \mathbb{R}_{\geq 0}$ and $k_i A_i$ are very ample for any $k_i > 0$ sufficiently divisible.

For the second statement, note that clearly $\sigma_P(\|D + \varepsilon A\|_S) \leq \sigma_P(\|D\|_S)$ for any prime divisor P on S. Let C be an effective $\mathbb{Q}$-divisor whose support does not contain S and $\delta > 0$ be a rational number such that $D \sim_{\mathbb{Q}} \delta A + C$. For $\varepsilon \geq 0$, we have

$$(1 + \varepsilon)\sigma_P(\|D\|_S) \leq \sigma_P(\|D + \varepsilon\delta A\|_S) + \varepsilon \operatorname{mult}_P(C|_S).$$

The assertion now follows by taking the limit as $\varepsilon \to 0^+$.

For the last statement, notice that the limits $\lim_{\varepsilon \to 0^+} \sigma_P(\|D + \varepsilon A\|_S)$ exist since $\sigma_P(\|D + \varepsilon A\|_S)$ is monotonic increasing and bounded. Here, the boundedness follows easily as

$$(D + \varepsilon A) \cdot S \cdot A^{\dim X - 2} \geq \sum \sigma_P(\|D + \varepsilon A\|_S) P \cdot (A|_S)^{\dim X - 2}.$$

Since for any ample $\mathbb{R}$-divisors A and A', there is a number $\delta > 0$ such that $A - \delta A'$ is ample, it follows from the first statement, that

$$\sigma_P(\|D + \varepsilon A\|_S) \leq \sigma_P(\|D + \varepsilon\delta A'\|_S)$$

and so that

$$\lim_{\varepsilon\to 0+}\sigma_P(\|D+\varepsilon A\|_S)\le\lim_{\varepsilon\to 0+}\sigma_P(\|D+\varepsilon A'\|_S).$$

By symmetry, we have the required equality. □

Definition 9.9. If D is a pseudo-effective $\mathbb{R}$-divisor such that $S\not\subset\mathbf{B}_-(D)$, then we define

$$N_\sigma(\|D\|_S)=\lim_{\varepsilon\to 0+}N_\sigma(\|D+\varepsilon A\|_S).$$

REMARK 9.10. Note that $N_\sigma(\|D\|_S)=\sum_{P\subset X}\sigma_P(\|D\|_S)P$ and that the set of prime divisors $P\subset S$ such that $\sigma_P(\|D\|_S)>0$ is countable.

REMARK 9.11. If $S\not\subset\mathbf{B}_+(D)$ and D' is a $\mathbb{Q}$-divisor such that $D'\equiv D$, then $S\not\subset\mathbf{B}_+(D')$ and $N_\sigma(\|D\|_S)=N_\sigma(\|D'\|_S)$.

This follows from the fact that if A is an ample $\mathbb{R}$-divisor, then $A'=A+D-D'$ is also ample. Therefore

$$N_\sigma(\|D\|_S)=\lim_{\varepsilon\to 0+}N_\sigma(\|D+\varepsilon A\|_S)=\lim_{\varepsilon\to 0+}N_\sigma(\|D'+\varepsilon A'\|_S)=N_\sigma(\|D'\|_S).$$

REMARK 9.12. Let $\mathcal{C}\subset N^1_{\mathbb{R}}(X)$ be the cone of $\mathbb{R}$-divisors on X such that S is not contained in $\mathbf{B}_+(D)$ for any $D\in\mathcal{C}$ and $\mathcal{C}'\subset N^1_{\mathbb{R}}(X)$ be the cone of $\mathbb{R}$-divisor on X such that S is not contained in $\mathbf{B}_-(D)$ for any $D\in\mathcal{C}$. Then $\mathcal{C}\subset\mathcal{C}'$ and for any prime divisor P on S, the functions $\sigma_P(\|\cdots\|_S)$ are continuous on $\mathcal{C}$ and lower semi-continuous on $\mathcal{C}'$.

Lemma 9.13. *Let D be a pseudo-effective $\mathbb{R}$-divisor such that S is not contained in $N_\sigma(D)$.*

Then S is not contained in $\mathbf{B}_-(D)$ and $D|_S-N$ is pseudo-effective for any divisor $N=\sum_{i=1}^k n_iP_i\le N_\sigma(\|D\|_S)$.

Proof. Let G be any $\mathbb{Q}$-divisor such that $G-D$ is ample. By (9.5), S is not contained in $\mathbf{B}(G)$. Since the restriction of $|G|_{\mathbb{R}}$ to S is not empty, we have that

$$G|_S-\sum_{i=1}^k\sigma_{P_i}(\|G\|_S)P_i\ge G|_S-N_\sigma(\|G\|_S)\ge 0.$$

Therefore $D|_S-\sum_{i=1}^k\sigma_{P_i}(\|D\|_S)P_i$ is a limit of effective divisors and hence is pseudo-effective. Since $N\le\sum_{i=1}^k\sigma_{P_i}(\|D\|_S)P_i$, the lemma follows. □

Lemma 9.14. *Let D be a pseudo-effective $\mathbb{R}$-divisor such that S is not contained in $N_\sigma(D)$ and A an ample $\mathbb{R}$-divisor such that $D+A$ is $\mathbb{Q}$-Cartier. Let P be a prime divisor on S. If $\sigma_P(\|D\|_S)>0$, then there is a positive integer l such that* $\operatorname{mult}_P\operatorname{Fix}(l(D+A))|_S<\sigma_P(\|D\|_S)$ *and if $\sigma_P(\|D\|_S)=0$, then there is a positive integer l such that* $\operatorname{mult}_P\operatorname{Fix}(l(D+A))|_S=0$.

Proof. Suppose that $\sigma_P(\|D\|_S) = 0$. Let H be a very ample divisor on X. Fix $l > 0$ such that $\frac{l}{2}A - (K_X + S) - (\dim X + 1)H$ is ample and let $\Delta \sim_{\mathbb{R}} D + A/2$ be an $\mathbb{R}$-divisor whose support does not contain S such that $\operatorname{mult}_P(\Delta|_S) < \frac{1}{l}$. Since $l(D + A) - (K_X + S + (\dim X + 1)H + l\Delta)$ is ample, $\mathcal{J}_{l\Delta|_S}(l(D + A))$ is globally generated cf. (6.26) and its sections lift to $H^0(X, \mathcal{O}_X(l(D + A)))$ cf. (6.6). Since $\operatorname{mult}_P(l\Delta|_S) < 1$, $\mathcal{J}_{l\Delta|_S}$ does not vanish along P and so $\operatorname{mult}_P \operatorname{Fix}((l(D + A))|_S) = 0$. The remaining assertion is similar. □

9.B Non-vanishing

Theorem 9.15. *Theorems 5.58, 5.59, 5.60 in dimensions $\leq n - 1$ and Theorems 5.56, 5.57 in dimensions $\leq n$ imply Theorem 5.58 in dimension n.*

Proof. **Step 1.** *We may assume that X is projective.*

Let F be the generic fiber of π, then F is projective over $\operatorname{Spec}(k)$ where $k = K(U)$. Let $\bar{k}$ be the algebraic closure of k so that $\bar{k}$ is isomorphic to $\mathbb{C}$. If $\bar{F} = F \times_{\operatorname{Spec}(k)} \operatorname{Spec}(\bar{k})$, then $\bar{F}$ is projective over $\operatorname{Spec}(\bar{k})$. Therefore, we may assume that there exists an effective $\mathbb{R}$-divisor $\bar{D}$ such that $K_{\bar{F}} + \Delta_{\bar{F}} \sim_{\mathbb{R}} \bar{D} \geq 0$. Equivalently, there are rational functions $\bar{f}_i \in K(\bar{F})$ and real numbers r_i such that $K_{\bar{F}} + \Delta_{\bar{F}} + \sum r_i(\bar{f}_i) \geq 0$. Notice that the $\bar{f}_i$ belong to a finite extension of $K(F)$ which we may assume to be Galois and that $K_{\bar{F}} + \Delta_{\bar{F}} = (K_X + \Delta)_{\bar{F}}$. It follows that there exist rational functions $f_i \in K(F)$ and real numbers t_i such that $(K_X + \Delta + \sum t_i(f_i))_{\bar{F}} \geq 0$. Taking the closure in X, there is an open subset $U^0 \subset U$ such that $K_X + \Delta + \sum t_i(f_i) \geq 0$ over U^0 and in fact we may choose an effective $\mathbb{R}$-divisor G on U such that $K_X + \Delta + \sum t_i(f_i) + \pi^*G \geq 0$. Therefore there exists an effective $\mathbb{R}$-divisor D such that $K_X + \Delta \sim_{\mathbb{R},U} D \geq 0$.

Step 2. *We may assume that X is smooth, $\Delta = A + B$ where A is a general ample $\mathbb{Q}$-divisor and* $\operatorname{Supp}(\Delta)$ *has simple normal crossings.*

Since Δ is big, after replacing it by an $\mathbb{R}$-linearly equivalent $\mathbb{R}$-divisor, we may assume that $\Delta = A + B$ where A is a general ample $\mathbb{Q}$-divisor. Let $f : Y \to X$ be a log resolution of (X, Δ) such that there is an effective and exceptional $\mathbb{Q}$-divisor F such that $f^*A - F$ is ample. Let $C \sim_{\mathbb{Q}} f^*A - F$ be a general ample $\mathbb{Q}$-divisor. We may write

$$K_Y + \Gamma = f^*(K_X + \Delta) + E$$

as in (3.21.1). We may assume that $(Y, \Gamma' = \Gamma - f^*A + F + C)$ is klt. Note that if $K_Y + \Gamma' \sim_{\mathbb{R},U} D \geq 0$, then $K_X + \Delta \sim_{\mathbb{R},U} f_*D \geq 0$.

Step 3. *We may assume that there is an integer $k > 0$ such that $h^0(\mathcal{O}_X(\lfloor mk(K_X + \Delta)\rfloor + kA))$ is an unbounded function of m.*

If this is not the case, then by (9.6), $K_X + \Delta \equiv N_\sigma(K_X + \Delta)$. Therefore $A'' = A + N_\sigma(K_X + \Delta) - (K_X + \Delta)$ is ample. Let $A' \sim_{\mathbb{R}} A''$ be an effective divisor such that $(X, \Delta' = A' + \Delta - A)$ is klt. We have that $K_X + \Delta' \sim_{\mathbb{R}} N_\sigma(K_X + \Delta) \geq 0$. By (5.57), there is an ltm $\phi : X \dashrightarrow Y$ for (X, Δ'). Since $K_X + \Delta \equiv K_X + \Delta'$ this is also an ltm for (X, Δ) cf. (5.39). By (5.10), $K_Y + \phi_*\Delta \sim_{\mathbb{R}} D' \geq 0$ and hence there is an effective $\mathbb{R}$ divisor D on X such that $K_X + \Delta \sim_{\mathbb{R}} D$.

Step 4. *We may assume that*

(9.15.1) (X, Δ) is plt and $\Delta = S + A + B$ where A is a general ample $\mathbb{Q}$-divisor and $S = \lfloor \Delta \rfloor$;

(9.15.2) $(S, \Delta_S + A|_S)$ is terminal where $(K_X + \Delta)|_S = K_S + \Delta_S$; and

(9.15.3) $\mathrm{Supp}(\Delta)$ and $\mathrm{Supp}(N_\sigma)$ have no common components.

Let $x \in X$ be a general point. By Step 3, we may assume that there are integers $k, m > 0$ and there is a divisor

$$H' \in |\lfloor mk(K_X + \Delta)\rfloor + kA| \qquad \text{such that} \qquad \mathrm{mult}_x(H') > kn.$$

It follows that there is an effective $\mathbb{R}$-divisor $H \sim_{\mathbb{R}} m(K_X + \Delta) + A$ with $\mathrm{mult}_x(H) > n$ and hence such that $(X, \Delta + H)$ is not lc at x. For any $t \geq 0$, we may write

$$\begin{aligned}(t+1)(K_X + \Delta) &= K_X + \frac{m-t}{m}A + B + t\left(K_X + \Delta + \frac{1}{m}A\right)\\ &\sim_{\mathbb{R}} K_X + \frac{m-t}{m}A + B + \frac{t}{m}H = K_X + \Delta_t.\end{aligned}$$

For $0 < \varepsilon \ll 1$, we let $A' = \frac{\varepsilon}{m}A$ and $u = m - \varepsilon$. It follows that

(9.15.1) (X, Δ_0) is klt;

(9.15.2) $\Delta_t \geq A'$ for any $t \in [0, u]$;

(9.15.3) x is contained in a non-klt center of (X, Δ_u).

Let $f : Y \to X$ be a log resolution of $(X, \Delta + H)$, then we may write

$$K_Y + \Gamma_t = \pi^*(K_X + \Delta_t) + E_t$$

as in (3.21.1). We may assume that there is an effective and exceptional $\mathbb{Q}$-divisor F such that $f^*A' - F$ is ample. Let $C \sim_{\mathbb{Q}} f^*A' - F$ be a general ample $\mathbb{Q}$-divisor and define

$$\Gamma'_t = \Gamma_t - f^*A' + F + C \sim_{\mathbb{Q}} \Gamma_t, \qquad \Gamma''_t = \Gamma'_t - (\Gamma'_t \wedge N_\sigma(K_Y + \Gamma'_t)).$$

The following properties now hold:

(9.15.1) (Y, Γ''_0) is klt;

(9.15.2) $\Gamma''_t \geq C$ for any $t \in [0, u]$;

(9.15.3) (Y, Γ''_u) is not lc;

(9.15.4) Y is smooth and Γ''_t has simple normal crossings support for any $t \in [0, u]$; and

(9.15.5) $\Gamma''_t \wedge N_\sigma(K_Y + \Gamma''_t) = 0$ for any $t \in [0, u]$.

Let $s = \sup\{t \in [0, u] | K_Y + \Gamma''_t \text{ is lc}\}$. Then (Y, Γ''_s) is dlt. Since C is a general ample $\mathbb{Q}$-divisor, it is easy to see that after replacing Γ''_s by an $\mathbb{R}$-equivalent $\mathbb{R}$-divisor, we may assume that (Y, Γ''_s) is plt. We may therefore assume that (X, Δ) satisfies properties (1) and (3) of Step 4. Property (2) may now be achieved as follows. Let $f : Y \to X$ be a log resolution of (X, Δ) such that there is an effective and exceptional $\mathbb{Q}$-divisor F such that $f^*A - F$ is ample. Let $C \sim_{\mathbb{Q}} f^*A - F$ be a general ample $\mathbb{Q}$-divisor. We may write

$$K_Y + \Gamma = f^*(K_X + \Delta) + E$$

as in (3.21.1). Let $T = f_*^{-1}S$. We may assume that $K_T + \Gamma_T = (K_Y + \Gamma)|_T$ is terminal. Let $\Gamma' = \Gamma - f^*A + F + C$. We may then also assume that $K_T + \Gamma'_T + C|_T = (K_Y + \Gamma' + C)|_T$ is terminal. After replacing Γ' by $\Gamma' - (\Gamma' \wedge N_\sigma(K_Y + \Gamma'))$, properties (1), (2) and (3) are satisfied.

Step 5. The theorem now follows from (9.16) below. □

Theorem 9.16. *Let $(X, \Delta = S + A + B)$ be a projective plt simple normal crossings pair where A is a general ample $\mathbb{Q}$-divisor, $B \geq 0$, $S = \lfloor \Delta \rfloor$ is irreducible and $(S, \Omega + A|_S)$ is terminal where $\Omega = (A + B)|_S$. If $K_X + \Delta$ is pseudo-effective and S is not contained in $N_\sigma(K_X + \Delta)$, then $K_X + \Delta \sim_{\mathbb{R}} D \geq 0$.*

Proof. Since $K_X + \Delta$ is pseudo-effective and S is not contained in $N_\sigma(K_X + \Delta)$, by (9.13), S is not contained in $\mathbf{B}_-(K_X + \Delta)$ and so $N_\sigma(\|K_X + \Delta\|_S)$ is defined. (Note that as observed above, the support of $N_\sigma(\|K_X + \Delta\|_S)$ consist of the union of countably many prime divisors in S.) Let

$$\Phi = \sum_{P \in \operatorname{Supp}(\Omega)} \sigma_P(\|K_X + \Delta\|_S)P \leq N_\sigma(\|K_X + \Delta\|_S)$$

and $\Theta = \Omega - \Omega \wedge \Phi$. By (9.13), $K_S + \Theta$ is pseudo-effective.

Let $W \subset \mathrm{WDiv}_{\mathbb{R}}(X)$ be the smallest rational affine subspace containing Δ, $V \subset \mathrm{WDiv}_{\mathbb{R}}(S)$ be the smallest rational affine subspace containing Θ and $U \subset \mathrm{WDiv}_{\mathbb{R}}(S)$ be the smallest rational affine subspace containing Φ.

There exists a number $1 \gg \varepsilon > 0$ such that $\varepsilon(K_X + \Delta) + \frac{1}{2}A$ is ample and if $\Delta' \in W$, $\|\Delta - \Delta'\| < \varepsilon$, then $\Delta' - \Delta + \frac{1}{2}A$ is ample.

By (5.60), there is a real number $\eta > 0$ and an integer $r > 0$ such that if $\Theta' \in V$ and $\|\Theta - \Theta'\| < \eta$, then $K_S + \Theta'$ is pseudo-effective and if $k > 0$ is an integer such that $k(K_S + \Theta')/r$ is Cartier, then $h^0(\mathscr{O}_S(k(K_S + \Theta'))) > 0$.

Let W' be the smallest rational affine subspace of $W \times U$ containing (Δ, Φ). In particular W' dominates W and U. By Diophantine Approximation cf. (7.2), there exist $\mathbb{Q}$-divisors $(\Delta_i, \Phi_i) \in W'$ and integers $k_i \gg 0$ such that:

(9.16.1) we may write $\Delta = \sum r_i \Delta_i$ where $r_i > 0$, $\sum r_i = 1$,

(9.16.2) $(X, \Delta_i = S + A + B_i)$ is plt and $(S, \Omega_i + A|_S)$ is terminal where $\Omega_i = (A + B_i)|_S$,

(9.16.3) $k_i \Delta_i / r$ are integral and $\|\Delta - \Delta_i\| < \varepsilon / k_i$,

(9.16.4) $k_i \Phi_i / r$ are integral and $\|\Phi - \Phi_i\| < \phi\varepsilon / k_i$ where ϕ is the minimum non-zero coefficient of Φ.

Claim 9.17. *We may assume that $\Theta_i = \Omega_i - (\Omega_i \wedge \Phi_i)$ belongs to V.*

Proof of Claim. Let P be a prime divisor on S contained in $\operatorname{Supp}(\Omega)$. Notice that if $\operatorname{mult}_P((\Delta-S)|_S) = \operatorname{mult}_P(\Phi)$, then $\operatorname{mult}_P((\Delta'-S)|_S) = \operatorname{mult}_P(\Phi')$ for any $(\Delta', \Phi') \in W'$. Let $(\Delta', \Phi') \in W'$. If $\|\Delta - \Delta'\| \ll 1$ and $\|\Phi - \Phi'\| \ll 1$, then

$$\operatorname{mult}_P((\Delta-S)|_S) > \operatorname{mult}_P(\Phi) \qquad \text{if and only if} \qquad \operatorname{mult}_P((\Delta'-S)|_S) > \operatorname{mult}_P(\Phi').$$

It follows that the assignment $\theta : W' \to \operatorname{WDiv}_{\mathbb{R}}(S)$ given by

$$\theta(\Delta', \Phi') = (\Delta' - S)|_S - \big((\Delta' - S)|_S \wedge \Phi'\big)$$

is rational linear and $\theta(\Delta, \Phi) = \Theta \in V$. Since V is a rational affine space, $\theta^{-1}(V) \subset W'$ is a rational affine subspace of W' containing (Δ, Φ) and so $\theta^{-1}(V) = W'$ and hence $V = \theta(W')$. □

Claim 9.18. *Let $P \in \operatorname{Supp}(\Omega)$, then for any $l > 0$ sufficiently divisible,*

$$\operatorname{mult}_P(\operatorname{Fix}|l(K_X + \Delta_i + \frac{1}{k_i}A)|_S) \leq l \operatorname{mult}_P(\Phi_i).$$

Proof of Claim. Since $\|\Phi - \Phi_i\| < \phi\varepsilon/k_i$, it suffices to show that

$$\operatorname{mult}_P(\operatorname{Fix}|l(K_X + \Delta_i + \frac{1}{k_i}A)|_S) \leq l \left(1 - \frac{\varepsilon}{k_i}\right) \operatorname{mult}_P(\Phi).$$

Since $\|\Delta_i - \Delta\| \ll 1$, $\varepsilon(K_X + \Delta_i) + \frac{1}{2}A$ is ample. Let $\eta > \varepsilon/k_i$ be a rational number such that $\eta(K_X + \Delta_i) + \frac{1}{2k_i}A$ is ample. Notice that

$$l\left(K_X + \Delta_i + \frac{1}{k_i}A\right) = l(1-\eta)\left(K_X + \Delta_i + \frac{1}{2k_i}A\right) + l\left(\eta(K_X + \Delta_i) + \frac{1+\eta}{2k_i}A\right).$$

Since $\Delta_i - \Delta + \frac{1}{2k_i}A$ is ample, we may assume that $\operatorname{Fix}|l(1-\eta)(K_X + \Delta_i + \frac{1}{2k_i}A)|_S \leq l(1-\eta)\Phi$ cf. (9.14). Since $\eta(K_X + \Delta_i) + \frac{1+\eta}{2k_i}A$ is ample, the assertion follows. □

Claim 9.19. *For all sufficiently divisible integers $m > 0$, we have*

$$\left|m\left(K_S + \Theta_i + \frac{1}{k_i}A\right)\right| + mF_i \subset \left|m(K_X + \Delta_i) + \frac{m}{k_i}A\right|_S,$$

where $F_i = \Omega_i - \Theta_i = \Omega_i \wedge \Phi_i$.

Proof of Claim. Immediate from (6.14). □

Claim 9.20. *We have*

$$|k_i(K_S + \Theta_i)| + k_i F_i \subset |k_i(K_X + \Delta_i)|_S.$$

Proof of claim. For any $\Sigma \in |k_i(K_S + \Theta_i)|$ and any $m > 0$ sufficiently divisible, we may choose a divisor $G \in |m(K_X + \Delta_i) + \frac{m}{k_i}A|$ such that $G|_S = \frac{m}{k_i}\Sigma + mF_i + \frac{m}{k_i}A|_S$. If we define $\Lambda = \frac{k_i-1}{m}G + B_i$, then

$$k_i(K_X + \Delta_i) \sim_{\mathbb{Q}} K_X + S + \Lambda + \frac{1}{k_i}A$$

where $\frac{1}{k_i}A$ is ample. By (6.6), we have a surjective homomorphism

$$H^0(X, \mathscr{J}_{S,\Lambda}(k_i(K_X + \Delta))) \to H^0(S, \mathscr{J}_{\Lambda|_S}(k_i(K_X + \Delta_i))).$$

(S, Ω_i) is klt, and so $\mathscr{J}_{\Omega_i} = \mathscr{O}_S$. Since

$$\Lambda|_S - (\Sigma + k_i F_i) = \left(\frac{k_i - 1}{m}G + B_i\right)\bigg|_S - (\Sigma + k_i F_i) \leq \Omega_i,$$

then by (2) of (6.4), we have $\mathscr{I}_{\Sigma+k_i F_i} \subset \mathscr{J}_{\Lambda|_S}$ and so

$$\Sigma + k_i F_i \in |k_i(K_X + \Delta_i)|_S$$

and the claim follows. □

Note that since $k_i(K_X + \Delta_i)/r$ and $k_i\Phi_i/r$ are Cartier, then so is $k_i(K_S + \Theta_i)/r$. As $\Theta_i \in V$ and $\|\Theta - \Theta_i\| < \eta$, $h^0(\mathscr{O}_S(k_i(K_S + \Theta_i))) > 0$ and so by (9.20), there exists an effective divisor D_i such that $k_i(K_X + \Delta_i) \sim_{\mathbb{Q}} D_i$. But then

$$K_X + \Delta \sim_{\mathbb{R}} \sum r_i D_i \geq 0. \qquad \square$$

REMARK 9.21. Related results may be found in [YT06], [Pău08], [Laz08].

EXERCISE 9.22. Let H and Δ be effective $\mathbb{R}$-divisors on a smooth variety X such that $\mathbf{B}(\Delta) = \emptyset$. Show that if $\Delta' \in |\Delta|_{\mathbb{R}}$ is general, then $\mathscr{J}(\Delta' + H) = \mathscr{J}(H)$.

EXERCISE 9.23. Let D be a pseudo-effective divisor on a normal projective variety. Show that if H is ample on X and $D \cdot H^{\dim(X)-1} \leq 0$, then $D \equiv 0$.

CHAPTER 10

Finiteness of log terminal models

Theorem 10.1. *Theorems 5.59, 5.60 in dimensions $\leq n-1$ and Theorems 5.56, 5.57, 5.58 in dimensions $\leq n$ imply Theorem 5.59 (1) in dimension n.*

Proof. We proceed by induction on the dimension of $\mathcal{C}$. The case when $\dim \mathcal{C} = 0$ follows immediately from Theorems 5.57 and 5.58 in dimension n.

Notice that $\mathrm{PSEF}(X/U)$ (the cone of pseudo-effective divisors over U) is closed. By compactness of $\mathcal{C} \cap \mathrm{PSEF}(X/U)$, it suffices to work locally around an $\mathbb{R}$-divisor $K_X + \Delta_0 \in \mathcal{C} \cap \mathrm{PSEF}(X/U)$. We are therefore free to replace $\mathcal{C}$ by an appropriate neighborhood of $K_X + \Delta_0$ inside $\mathcal{C}$.

Step 1. *We may assume that $K_X + \Delta_0$ is nef over U.*

By Theorems 5.57 and 5.58 in dimension n, there exists an ltm $\phi : X \dashrightarrow Y$ for $K_X + \Delta_0$ over U. Note that if $K_X + \Delta$ is pseudo-effective over U, then $K_Y + \phi_*\Delta$ is pseudo-effective over U. After shrinking $\mathcal{C}$, we have for all $K_X + \Delta \in \mathcal{C}$ that:

(10.1.1) $K_Y + \phi_*\Delta$ is klt cf. (10.5), and

(10.1.2) $a(F; X, \Delta) < a(F; Y, \phi_*\Delta)$ for any ϕ-exceptional divisor $F \subset X$.

It follows that if $\psi : Y \dashrightarrow Z$ is an ltm for $(Y, \phi_*\Delta)$ over U, then $\psi \circ \phi : X \dashrightarrow Z$ is an ltm for (X, Δ) over U cf. (10.6). Replacing X by Z, Δ by $\phi_*\Delta$ and $\mathcal{C}$ by $\phi_*\mathcal{C} = \{\phi_*(K_X + \Delta) | K_X + \Delta \in \mathcal{C}\}$, we may therefore assume that $K_X + \Delta_0$ is nef over U.

Step 2. *We may assume that $K_X + \Delta_0 \sim_{\mathbb{R},U} 0$.*

Since $K_X + \Delta_0$ is nef over U and Δ_0 is big over U, by the Basepoint-free theorem (cf. (5.10)), $K_X + \Delta_0$ is semiample over U and so there is a morphism $\psi : X \to Z$ over U and an $\mathbb{R}$-divisor H on Z ample over U such that $K_X + \Delta_0 \sim_{\mathbb{R},U} \psi^* H$.

After further shrinking $\mathcal{C}$, we may assume that any given ltm for (X, Δ) over Z is an ltm for (X, Δ) over U cf. (10.7). We may hence replace U by Z so that we may assume that $K_X + \Delta_0 \sim_{\mathbb{R},U} 0$.

Step 3. Pick any $K_X + \Theta \in \mathcal{C}$ with $\Theta \neq \Delta_0$ and let $K_X + \Delta$ be the point on the boundary of $\mathcal{C}$ such that

$$\Theta - \Delta_0 = \lambda(\Delta - \Delta_0) \qquad 0 < \lambda \leq 1.$$

Then we have that

$$K_X + \Theta = \lambda(K_X + \Delta) + (1 - \lambda)(K_X + \Delta_0) \sim_{\mathbb{R},U} \lambda(K_X + \Delta).$$

It follows that $K_X + \Delta$ is pseudo-effective over U if and only if $K_X + \Theta$ is pseudo-effective over U and $\phi : X \dashrightarrow Y$ is an ltm for (X, Δ) over U if and only if it is an ltm for (X, Θ) over U cf. (5.40). Since the boundary of $\mathcal{C}$ is a rational polytope of strictly smaller dimension, the induction is now complete. □

Theorem 10.2. *Theorems 5.59, 5.60 in dimensions $\leq n - 1$ and Theorems 5.56, 5.57, 5.58 in dimensions $\leq n$ imply Theorem 5.59 (2) in dimension n.*

Proof. Let $\nu : X' \to X$ be a $\mathbb{Q}$-factorialization of X cf. (5.41) so that X' is $\mathbb{Q}$-factorial and ν is a small morphism. If (X, Δ) is klt, then $\phi : X \dashrightarrow Y$ is an ltm of (X, Δ) if and only if $\phi \circ \nu : X' \dashrightarrow Y$ is an ltm of $(X', \nu_*^{-1}\Delta)$. Replacing X by X' we may assume that X is $\mathbb{Q}$-factorial.

Notice that if $\phi : X \dashrightarrow Y$ is any ltm of (X, Δ) over U and $H \in \mathrm{WDiv}_{\mathbb{R}}(Y)$ is ample over U, then ϕ is an ltm over U for any klt pair (X, Δ') such that $\Delta' \equiv_U \Delta + \varepsilon\phi_*^{-1}H$ and $0 < \varepsilon \ll 1$. Since $K_Y + \phi_*\Delta'$ is ample over U, ϕ is in fact the unique ltm over U for (X, Δ').

It suffices therefore to show that there exists a rational polytope $\mathcal{C} \subset \mathcal{C}'$ contained in a finite dimensional subspace of $\mathrm{WDiv}_{\mathbb{R}}(X)$ such that if $K_X + \Delta \in \mathcal{C}$ and H' is an $\mathbb{R}$-Cartier divisor on X, then there exists a klt pair $K_X + \Delta' \in \mathcal{C}'$ and a constant $\varepsilon > 0$ such that $\Delta' \equiv_U \Delta + \varepsilon H'$. This can be achieved as follows: We may assume that $\mathcal{C}$ is a simplex (i.e., that $\mathcal{C}$ has $n + 1$ vertices where n is the dimension of the affine subspace spanned by $\mathcal{C}$). By (10.8), we may assume that there is a divisor A ample over U such that $\Delta \geq A$ for all $K_X + \Delta \in \mathcal{C}$. We may now translate $\mathcal{C}$ by $A' + B' - A$ where A' is a general ample $\mathbb{Q}$-divisor over U, $B' \geq 0$ is a $\mathbb{Q}$-divisor such that $A \sim_{\mathbb{R},U} A' + B'$, the components of $\mathrm{Supp}(B')$ generate $\mathrm{WDiv}_{\mathbb{R}}(X)/ \equiv_U$, and for any $K_X + \Delta \in \mathcal{C}$, $\Delta - A + A' + B'$ is effective and $(X, \Delta - A + A' + B')$ is klt. We then let

$$\mathcal{C}' = \{K_X + \Delta' = K_X + \Delta + C \mid K_X + \Delta \in \mathcal{C},\ \mathrm{Supp}(C) \subset \mathrm{Supp}(B') \text{ and } ||C|| < \varepsilon\}$$

where $0 < \varepsilon \ll 1$. □

Theorem 10.3. *Theorem 5.60 in dimensions $\leq n - 1$ and Theorems 5.56, 5.57, 5.58, 5.59 in dimensions $\leq n$ imply Theorem 5.60 in dimension n.*

Proof. (1) is immediate from (5.58) and (5.10). We now prove (2). Notice that if $K_X + \Delta'$ is not pseudo-effective, then $\mathbf{B}(K_X + \Delta') = X$. Therefore, we may assume that $K_X + \Delta'$ is pseudo-effective. Let $\phi : X \dashrightarrow Y$ be an ltm for $K_X + \Delta$ over Z. We may pick $0 < \delta$ such that if $||\Delta - \Delta'|| \leq \delta$, then $K_Y + \phi_*\Delta'$ is klt cf. (10.5) and $a(F; X, \Delta') < a(F; Y, \phi_*\Delta')$ for any ϕ-exceptional divisor $F \subset X$. By (10.6), if $\psi : Y \dashrightarrow Z$ is an ltm for $K_Y + \phi_*\Delta'$ over Z, then $\psi \circ \phi$ is an ltm for $K_X + \Delta'$ over Z. By (5.36), all ϕ-exceptional divisors are contained in $\mathbf{B}(K_X + \Delta')$.

We now prove (3). Let $\phi : X \dashrightarrow Y$ be an ltm for $K_X + \Delta$ over Z and let $\mathcal{C}$ be the set of all $\Delta' \in W$ such that $||\Delta - \Delta'|| \leq \eta$ and ϕ is an ltm for $K_X + \Delta'$ over Z. Since Δ is big, we may assume that there is $\mathbb{Q}$-divisor A on Y ample over Z such that if $||\Delta - \Delta'|| \leq \eta$, then $\phi_*\Delta \geq A$. If η is sufficiently small, then $\phi_*(\Delta' - \Delta) + A/2$ is ample over Z for any $\Delta' \in W$ such that $\|\Delta - \Delta'\| \leq \eta$ and moreover

$$\mathcal{C} = \{\Delta' \in W \mid K_Y + \phi_*\Delta' \text{ is nef over } Z, \text{ and } ||\Delta - \Delta'|| \leq \eta\}.$$

By the Cone theorem cf. (5.4), there are finitely many negative extremal rays $R_1, \ldots, R_k$ for $K_Y + \phi_*\Delta - A/2$ over Z. Since

$$K_Y + \phi_*\Delta' = (K_Y + \phi_*\Delta - A/2) + \phi_*(\Delta' - \Delta) + A/2,$$

if $K_Y + \phi_*\Delta'$ is not nef over Z, then there is an integer $1 \leq i \leq k$ such that R_i is a negative extremal ray for $K_Y + \phi_*\Delta - A/2$ over Z. Therefore $\mathcal{C}$ is cut out by a subset of the finite collection of extremal rays $R_i = \mathbb{R}^+[C_i]$ and hence $\mathcal{C}$ is a rational polytope. Therefore the affine subspace of $\mathrm{WDiv}_{\mathbb{R}}(X)$ generated by $\mathcal{C}$ is W. We may then assume that $\mathcal{C} = \{\Delta' \in W : ||\Delta - \Delta'|| \leq \eta\}$.

Since Y is $\mathbb{Q}$-factorial, there is an integer $l > 0$ such that if G is an integral Weil divisor, then lG is Cartier. By (5.3), there exists an integer $r > 0$ such that if $k(K_Y + \phi_*\Delta')/r$ is integral, then $k(K_Y + \phi_*\Delta')$ is generated by global sections. Let $p : W \to X$ and $q : W \to Y$ be a common resolution, then by definition of ltm, $p^*(K_X + \Delta') = q^*(K_Y + \phi_*\Delta') + E$ where $E \geq 0$ is q-exceptional and the support of $p_*(E)$ is the union of all ϕ-exceptional divisors on X. It follows that the support of $\mathrm{Fix}(k(K_X + \Delta'))$ is given by the divisors contained in $\mathbf{B}(K_X + \Delta')$, i.e., by the divisors contained in the support of $p_*(E)$. Since ϕ is an ltm for $K_X + \Delta$ over Z, these divisors are also contained in $\mathbf{B}(K_X + \Delta)$. □

REMARK 10.4. Some related ideas and techniques may be found in [Sho96].

EXERCISE 10.5. Let (X, Δ_0) be a $\mathbb{Q}$-factorial klt pair and $\Delta_0 \in V \subset \mathrm{WDiv}_{\mathbb{R}}(X)$ a finite dimensional vector space. Show that if there exists an $\varepsilon > 0$ such that $0 \leq \Delta \in V$ and $||\Delta - \Delta_0|| \leq \varepsilon$, then (X, Δ) is klt.

EXERCISE 10.6. Let (X, Δ) be a $\mathbb{Q}$-factorial klt pair, $\pi : X \to U$ a projective morphism, $\phi : X \dashrightarrow Y$ a rational map over U that extracts no divisors such that Y is normal, $\mathbb{Q}$-factorial and projective over U. If $a(F; X, \Delta) < a(F; Y, \phi_*\Delta)$ for any ϕ-exceptional divisor $F \subset X$ and if $\psi : Y \dashrightarrow Z$ is an ltm of $(Y, \phi_*\Delta)$ over U, then $\psi \circ \phi$ is an ltm of (X, Δ) over U.

EXERCISE 10.7. Let $\psi : X \to Z$ and $\eta : Z \to U$ be projective morphisms of normal varieties and $\phi : X \dashrightarrow Y$ be a birational map over Z. If H is an ample $\mathbb{R}$-divisor on Z and (X, Δ_0) is a $\mathbb{Q}$-factorial klt pair such that Δ_0 is big over U and $K_X + \Delta_0 \sim_{\mathbb{R},U} \psi^* H$, show that there exists $0 < \varepsilon \ll 1$ such that if $||\Delta - \Delta_0|| \leq \varepsilon$, then ϕ is an ltm for (X, Δ) over Z if and only if it is an ltm for (X, Δ) over U.

EXERCISE 10.8. Let $\pi : X \to U$ be a projective morphism and A a π-ample divisor on X. Let $\mathcal{P} \subset \mathrm{WDiv}_{\mathbb{R}}(X)$ be a simplex such that if $\Delta \in \mathcal{P}$, then Δ is π-big and $K_X + \Delta$ is klt. Show that there exists an $\varepsilon > 0$ and a linear map $L : \mathrm{WDiv}_{\mathbb{R}}(X) \to \mathrm{WDiv}_{\mathbb{R}}(X)$ such that $L(\Delta) \sim_{U,\mathbb{Q}} \Delta$ for any $\Delta \in \mathcal{P}$, $K_X + L(\Delta)$ is klt and $L(\Delta) - \varepsilon A \geq 0$.

Part III

Compact moduli spaces of canonically polarized varieties

CHAPTER 11

Moduli problems

11.A Representing functors

Let Sets denote the category of sets and Cat an arbitrary category. Further let

$$\mathscr{F} : \mathsf{Cat} \to \mathsf{Sets}$$

be a contravariant functor. Recall that $\mathscr{F}$ is **representable** if there is an object $\mathfrak{M} \in \operatorname{Ob}\mathsf{Cat}$ such that $\mathscr{F} \simeq \operatorname{Hom}_{\mathsf{Cat}}(\_\_, \mathfrak{M})$. If such an $\mathfrak{M}$ exists, it is called a **universal object** or a **fine moduli space** for $\mathscr{F}$.

11.B Moduli functors

NOTATION 11.1. Let $f : X \to B$ be a morphism and $\mathscr{K}$ and $\mathscr{L}$ two line bundles on X. Then

$$\mathscr{K} \sim_B \mathscr{L}$$

will mean that there exists a line bundle $\mathscr{N}$ on B such that $\mathscr{K} \simeq \mathscr{L} \otimes f^*\mathscr{N}$.

EXERCISE 11.2. Prove that if B and X_b for all $b \in B$ are integral of finite type and f is flat and projective, then $\mathscr{K} \sim_B \mathscr{L}$ is equivalent to the condition that $\mathscr{K}\big|_{X_b} \simeq \mathscr{L}\big|_{X_b}$ for all $b \in B$.

DEFINITION 11.3. Let S be a scheme and Sch_S the category of S-schemes. Let

$$\mathcal{MP} : \mathsf{Sch}_S \to \mathsf{Sets}$$

be the **moduli functor of polarized proper schemes over** S:

(11.3.1) For an object $B \in \operatorname{Ob} \mathsf{Sch}_S$,

$$\mathcal{MP}(B) := \{(f : X \to B, \mathcal{L}) \mid f \text{ is a flat, projective morphism and } \mathcal{L} \text{ is an } f\text{-ample line bundle on } X\}/\simeq$$

where "$\simeq$" is defined as follows: $(f_1 : X_1 \to B, \mathcal{L}_1) \simeq (f_2 : X_2 \to B, \mathcal{L}_2)$ if and only if there exists a B-isomorphism $\phi : X_1/B \xrightarrow{\simeq} X_2/B$ such that $\mathcal{L}_1 \sim_B \phi^*\mathcal{L}_2$.

(11.3.2) For a morphism $\alpha \in \operatorname{Hom}_{\mathsf{Sch}_S}(A, B)$,

$$\mathcal{MP}(\alpha) := (\_) \times_B \alpha,$$

i.e.,

$$\begin{aligned} \mathcal{MP}(\alpha) : \mathcal{MP}(B) &\longrightarrow \mathcal{MP}(A) \\ (f : X \to B, \mathcal{L}) &\longmapsto (f_A : X_A \to A, \mathcal{L}_A). \end{aligned}$$

REMARK 11.4. The above definition has the disadvantage that it does not satisfy faithfully flat descent, cf. [BLR90, 6.1]. This is essentially caused by similar problems with the naive definition of the relative Picard functor [Gro62, 232] or [BLR90, 8.1]. The problem may be dealt with by appropriate sheafification of $\mathcal{MP}$. The notion of canonical polarization below also provides a natural solution in many cases. For details see [Vie95, §1].

Considering our current aim, we leave these worries behind, but warn the reader that they should be addressed.

In any case, unfortunately, the functor $\mathcal{MP}$ is too big to handle, so we need to study some of its subfunctors that are more accessible. One reason is that $\mathcal{MP}$ does not take into account any discrete invariants. If we follow our original plan for classification and start by fixing certain discrete invariants, then we are led to study natural subfunctors of $\mathcal{MP}$.

DEFINITION 11.5. Let k be an algebraically closed field of characteristic 0 and Sch_k the category of k-schemes. Let $h \in \mathbb{Q}[t]$ and $\mathcal{M}_h^{\mathrm{smooth}} : \mathsf{Sch}_k \to \mathsf{Sets}$ the following functor:

(11.5.1) For an object $B \in \operatorname{Ob} \mathsf{Sch}_k$,

$$\mathcal{M}_h^{\mathrm{smooth}}(B) := \{f : X \to B \mid f \text{ is a smooth projective family such that } \forall b \in B, \omega_{X_b} \text{ is ample and } \chi(X_b, \omega_{X_b}^{\otimes m}) = h(m)\}/\simeq$$

where "$\simeq$" is defined as follows: $(f_1 : X_1 \to B) \simeq (f_2 : X_2 \to B)$ if and only if there exists a B-isomorphism $\phi : X_1/B \xrightarrow{\simeq} X_2/B$.

(11.5.2) For a morphism $\alpha \in \operatorname{Hom}_{\mathsf{Sch}_k}(A, B)$,

$$\mathcal{M}_h^{\mathrm{smooth}}(\alpha) := (\_) \times_B \alpha.$$

REMARK 11.6. For $S = \operatorname{Spec} k$, $\mathcal{M}_h^{\text{smooth}}$ is a subfunctor of $\mathcal{MP}$.

REMARK 11.7. Recall that by [Har77, III.10.1(b)], being smooth is invariant under base change.

EXAMPLE 11.8.

$$\mathcal{M}_h^{\text{smooth}}(\operatorname{Spec} k) = \{X \mid X \text{ is a smooth projective variety with } \omega_X \text{ ample and } \chi(\omega_X^{\otimes m}) = h(m)\}.$$

QUESTION 11.9. What does it mean that $\mathcal{M}_h^{\text{smooth}}$ is representable?

OBSERVATIONS 11.10. Suppose(!) that $\mathcal{M}_h^{\text{smooth}}$ is representable, i.e., assume (but do not believe) that there exists an $\mathfrak{M} \in \operatorname{Ob} \mathsf{Sch}_k$ such that $\mathcal{M}_h^{\text{smooth}} \simeq \operatorname{Hom}_{\mathsf{Sch}_k}(\_\_, \mathfrak{M})$. Then one makes the following observations.

(11.10.1) First let $B = \operatorname{Spec} k$. Then $\mathcal{M}_h^{\text{smooth}}(\operatorname{Spec} k) \simeq \operatorname{Hom}_{\mathsf{Sch}_k}(\operatorname{Spec} k, \mathfrak{M}) = \mathfrak{M}(k)$, the set of k-points of $\mathfrak{M}$. In other words, the set of closed points of $\mathfrak{M}$ are in one-to-one correspondence with smooth projective varieties X with ω_X ample and $\chi(X, \omega_X^{\otimes m}) = h(m)$. For such a variety X, its corresponding point in $\mathfrak{M}(k)$ will be denoted by $[X]$.

(11.10.2) Next let $B = \mathfrak{M}$. Then one obtains that $\mathcal{M}_h^{\text{smooth}}(\mathfrak{M}) \simeq \operatorname{Hom}_{\mathsf{Sch}_k}(\mathfrak{M}, \mathfrak{M})$. Now let $(\mathfrak{f} : \mathfrak{U} \to \mathfrak{M}) \in \mathcal{M}_h^{\text{smooth}}(\mathfrak{M})$ be the element corresponding to the identity $\operatorname{id}_{\mathfrak{M}} \in \operatorname{Hom}_{\mathsf{Sch}_k}(\mathfrak{M}, \mathfrak{M})$. For a closed point $x : \operatorname{Spec} k \to \mathfrak{M}$ one has by functoriality that $x = [\mathfrak{U}_x]$, where $\mathfrak{U}_x = \mathfrak{U} \times_{\mathfrak{M}} x$. Therefore, $(\mathfrak{f} : \mathfrak{U} \to \mathfrak{M})$ is a **tautological family**.

(11.10.3) Finally, let B be arbitrary. Then by the definition of representability one has that $\mathcal{M}_h^{\text{smooth}}(B) \simeq \operatorname{Hom}_{\mathsf{Sch}_k}(B, \mathfrak{M})$, i.e., every family $(f : X \to B) \in \mathcal{M}_h^{\text{smooth}}(B)$ corresponds in a one-to-one manner to a morphism $\mu_f : B \to \mathfrak{M}$. Applying the functor $\mathcal{M}_h^{\text{smooth}}(\_\_) \simeq \operatorname{Hom}_{\mathsf{Sch}_k}(\_\_, \mathfrak{M})$ to μ_f leads to the following:

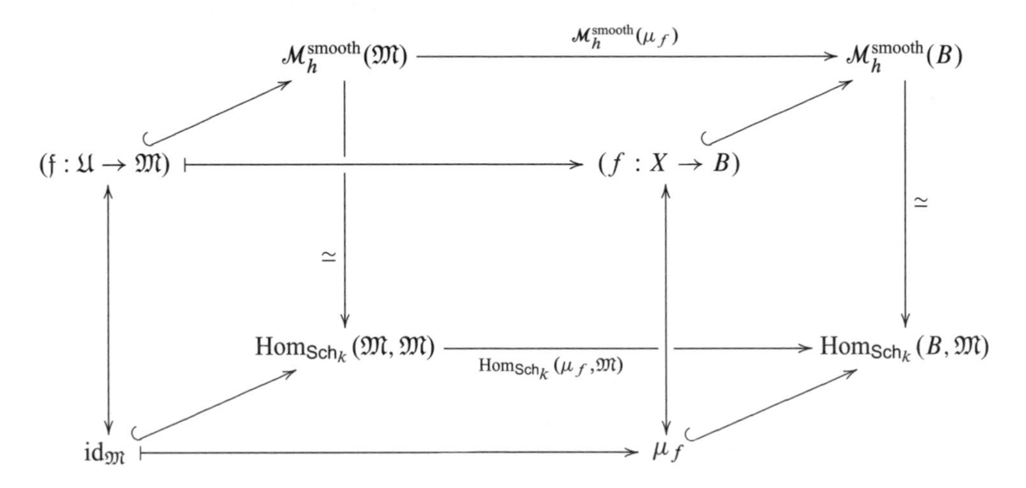

By (11.5.2) this implies that $(f : X \to B) \simeq (\mathfrak{f} \times_{\mathfrak{M}} \mu_f : \mathfrak{U} \times_{\mathfrak{M}} B \to B)$, so $(\mathfrak{f} : \mathfrak{U} \to \mathfrak{M})$ is actually a **universal family**.

(11.10.4) Let $(f : X \to B) \in \mathcal{M}_h^{\text{smooth}}(B)$ be a non-trivial family, all of whose members are isomorphic. For an example of such a family see (11.12) below. Let F denote the variety to which the fibers of f are isomorphic, i.e., $F \simeq X_b$ for all $b \in B$. Then by (11.10.2) $\mu_f(b) = [F] \in \mathfrak{M}$ for all $b \in B$. However, for this f then $(\mathfrak{f} \times_{\mathfrak{M}} \mu_f : \mathfrak{U} \times_{\mathfrak{M}} B \to B) \simeq (B \times F \to B)$, which is a contradiction.

CONCLUSION 11.11. Our original assumption led to a contradiction, so we have to conclude that $\mathcal{M}_h^{\text{smooth}}$ is *not* representable.

EXERCISE 11.12. Let B and C be two smooth projective curves admitting non-trivial unramified double covers $\widetilde{B} \to B \simeq \widetilde{B} \big/ \mathbb{Z}_2$ and $\widetilde{C} \to C \simeq \widetilde{C} \big/ \mathbb{Z}_2$. Consider the diagonal $\mathbb{Z}_2$-action on $\widetilde{B} \times \widetilde{C}$: $\sigma(b,c) := (\sigma(b), \sigma(c))$ for $\sigma \in \mathbb{Z}_2$ and let $X = \widetilde{B} \times \widetilde{C} \big/ \mathbb{Z}_2$ and $f : X \to B$ the induced morphism $[(b,c) \sim (\sigma(b), \sigma(c))] \mapsto [b \sim \sigma(b)]$. Prove that the fibers of f are all isomorphic to $\widetilde{C}$, but $X \not\simeq B \times \widetilde{C}$. Similar examples may be constructed as soon as there exists a non-trivial representation $\pi_1(B) \to \operatorname{Aut} C$.

11.C Coarse moduli spaces

Since we cannot expect our moduli functors to be representable, we have to make do with something weaker.

DEFINITION 11.13. A functor $\mathcal{F} : \mathsf{Sch}_k \to \mathsf{Sets}$ is **coarsely representable** if there exists an $\mathfrak{M} \in \operatorname{Ob} \mathsf{Sch}_k$ and a natural transformation

$$\eta : \mathcal{F} \to \operatorname{Hom}_{\mathsf{Sch}_k}(\_\_, \mathfrak{M})$$

such that

(11.13.1) $\eta_{\operatorname{Spec} k} : \mathcal{F}(\operatorname{Spec} k) \xrightarrow{\simeq} \operatorname{Hom}_{\mathsf{Sch}_k}(\operatorname{Spec} k, \mathfrak{M}) = \mathfrak{M}(k)$ is an isomorphism, and

(11.13.2) given an arbitrary $\mathfrak{N} \in \operatorname{Ob} \mathsf{Sch}_k$ and a natural transformation

$$\zeta : \mathcal{F} \to \operatorname{Hom}_{\mathsf{Sch}_k}(\_\_, \mathfrak{N}),$$

there exists a unique natural transformation

$$\nu : \operatorname{Hom}_{\mathsf{Sch}_k}(\_\_, \mathfrak{M}) \to \operatorname{Hom}_{\mathsf{Sch}_k}(\_\_, \mathfrak{N})$$

such that

$$\nu \circ \eta = \zeta.$$

If such an $\mathfrak{M}$ exists, it is called a **coarse moduli space** for $\mathcal{F}$.

Let us now reconsider the question and observations we made in (11.9) and (11.10) with regard to this new definition.

QUESTION 11.14. What does it mean that $\mathcal{M}_h^{\text{smooth}}$ is coarsely representable?

OBSERVATIONS 11.15. Assume that there exists an $\mathfrak{M}_h \in \operatorname{Ob} \mathsf{Sch}_k$ satisfying the conditions listed in Definition 11.13 above, i.e., assume that $\mathcal{M}_h^{\text{smooth}}$ is coarsely represented by $\mathfrak{M}_h$. Then one makes the following observations.

(11.15.1) Let $B = \operatorname{Spec} k$. Then by (11.13.1) we still have $\mathcal{M}_h^{\text{smooth}}(\operatorname{Spec} k) \simeq \mathfrak{M}_h(k)$, the set of k-points of $\mathfrak{M}_h$. In other words, the set of closed points of $\mathfrak{M}_h$ are in one-to-one correspondence with smooth projective varieties X with ω_X ample and $\chi(X, \omega_X^{\otimes m}) = h(m)$. For such a variety, X its corresponding point in $\mathfrak{M}_h(k)$ will be denoted by $[X]$ (again).

(11.15.2) Let $B = \mathfrak{M}_h$. Then there exists a map

$$\eta_{\mathfrak{M}_h} : \mathcal{M}_h^{\text{smooth}}(\mathfrak{M}_h) \to \operatorname{Hom}_{\mathsf{Sch}_k}(\mathfrak{M}_h, \mathfrak{M}_h),$$

but there is no guarantee that $\operatorname{id}_{\mathfrak{M}_h} \in \operatorname{Hom}_{\mathsf{Sch}_k}(\mathfrak{M}_h, \mathfrak{M}_h)$ is in the image of $\eta_{\mathfrak{M}_h}$, and hence a tautological family $(\mathfrak{f}_h : \mathfrak{U}_h \to \mathfrak{M}_h)$ may not exist.

(11.15.3) Let B be arbitrary. Then there exists a map $\eta_B : \mathcal{M}_h^{\text{smooth}}(B) \to \operatorname{Hom}_{\mathsf{Sch}_k}(B, \mathfrak{M})$, i.e., every family $(f : X \to B) \in \mathcal{M}_h^{\text{smooth}}(B)$ corresponds to a morphism $\mu_f : B \to \mathfrak{M}$, which still has some useful properties. Since it is given by a natural transformation, we have that for all $b \in B$,

$$\mu_f(b) = [X_b].$$

Applying the functors $\mathcal{M}_h^{\text{smooth}}(\_\_)$ and $\operatorname{Hom}_{\mathsf{Sch}_k}(\_\_, \mathfrak{M})$ to μ_f leads to the following:

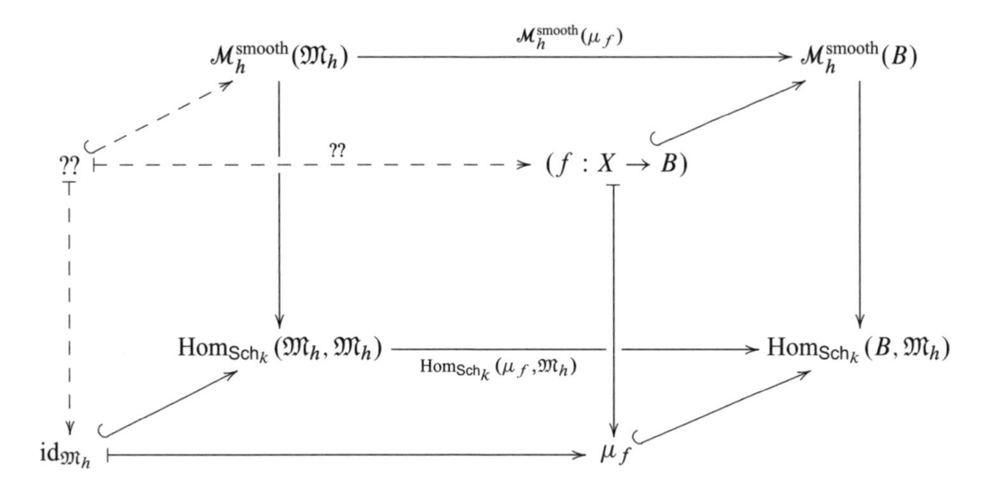

Figure 11.15.1:

We have observed in (11.15.2) that there may not be a tautological family

$$(\mathfrak{f}_h : \mathfrak{U}_h \to \mathfrak{M}_h) \in \mathcal{M}_h^{\text{smooth}}(\mathfrak{M}_h)$$

that maps to $\mathrm{id}_{\mathfrak{M}_h}$. However, even if such a family existed, we could not conclude that it maps to $(f : X \to B)$ via $\mathcal{M}_h^{\text{smooth}}(\mu_f)$, because the vertical arrows in Figure 11.15.1 are not necessarily one-to-one. In other words, even if we find a tautological family, it is not necessarily a **universal family**.

(11.15.4) Finally, let $(f : X \to B) \in \mathcal{M}_h^{\text{smooth}}(B)$ be a non-trivial family all of whose members are isomorphic. Let F denote the fiber of f, i.e., $F \simeq X_b$ for all $b \in B$. Then by (11.15.3) $\mu_f(b) = [F] \in \mathfrak{M}$ for all $b \in B$. However, this does not lead to a contradiction now (see the remark at the end of (11.15.3)).

CHAPTER 12

Hilbert schemes

In the previous chapter we saw that moduli functors are usually not representable. In this chapter we will see examples of representable functors.

We will omit some of the proofs. The reader is encouraged to read the much more detailed and complete [Kol96, Chapter I].

12.A The Grassmannian functor

Throughout this section we will use the notation for the pull-back of a sheaf introduced in (2.15).

Let S be a scheme, $r \in \mathbb{N}$, and $\mathscr{E}$ a locally free sheaf on S. With a slight abuse of terminology we will call a locally free subsheaf $\mathscr{F} \hookrightarrow \mathscr{E}$ a **subbundle** if the cokernel of the embedding, $\mathscr{E}/\mathscr{F}$ is locally free. If the rank of $\mathscr{F}$ is 1 we call it a **line subbundle**.

We define the **Grassmannian functor**,

$$\mathscr{G}\!\mathit{rass}_S(r,\mathscr{E}) : \mathsf{Sch}_S \to \mathsf{Sets}$$

as follows. For any $Z \in \operatorname{Ob} \mathsf{Sch}_S$,

$$\mathscr{G}\!\mathit{rass}_S(r,\mathscr{E})(Z) := \left\{\mathscr{V} \mid \mathscr{V} \subseteq \mathscr{E}_Z \text{ is a subbundle of rank } r\right\}$$

and for any $\phi \in \operatorname{Hom}_{\mathsf{Sch}_S}(Y, Z)$,

$$\begin{aligned} \mathscr{G}\!\mathit{rass}_S(r,\mathscr{E})(\phi) : \mathscr{G}\!\mathit{rass}_S(r,\mathscr{E})(Z) &\to \mathscr{G}\!\mathit{rass}_S(r,\mathscr{E})(Y) \\ \mathscr{V} &\mapsto \mathscr{V}_Y \end{aligned}$$

Theorem 12.1. *The functor $\mathcal{G}rass_S(r, \mathscr{E})$ is represented by a projective S-scheme* $\mathrm{Grass}_S(r, \mathscr{E})$ *and a (universal) subbundle $\mathscr{U} \subseteq \mathscr{E}_{\mathrm{Grass}_S(r,\mathscr{E})}$ of rank r.*

We will illustrate how one constructs a scheme representing a functor on this relatively simple example. It will also demonstrate that being "simple" is indeed relative.

Before we do this, we need a few auxiliary preparatory statements.

DEFINITION 12.2. Let V be a vector space, $p, q \in \mathbb{N}$, and for any $v \in \wedge^p V$ define a linear map by the following formula:

$$\begin{aligned} \wedge_v : \wedge^q V &\to \wedge^{p+q} V \\ x &\mapsto v \wedge x. \end{aligned}$$

Next let $\Xi \subseteq \wedge^p V$ be a subspace and set

$$\ker^q \Xi := \bigcap_{\xi \in \Xi} \ker \wedge_\xi \subseteq \wedge^q V.$$

EXERCISE 12.3. Prove that $\ker^q$ is contravariant on the set of subspaces of $\wedge^p V$: if $\Xi \subseteq \Sigma \subseteq \wedge^p V$, then $\ker^q \Sigma \subseteq \ker^q \Xi$. Also prove that $\Xi \subseteq \ker^p(\ker^q \Xi)$.

Lemma 12.4. *Let $W \subseteq V$ be a subspace of dimension $p + q - 1$. Then*

(12.4.1) $\ker^p(\wedge^q W) = \wedge^p W$.

Further let $0 \neq \Xi \subseteq \wedge^p V$ be a non-zero subspace such that $\ker^q \Xi \supseteq \wedge^q W$. *Then, as subspaces of $\wedge^p V$,*

(12.4.2) $\Xi \subseteq \wedge^p W$. In particular, if $q = 1$, then $\Xi = \wedge^p W$.

(12.4.3) If $q = 1$, then $\dim \ker^q \Xi \leq p$.

Proof. Writing V as a direct sum $V = W \oplus W'$ and expressing $\wedge^q V$ in terms of exterior powers of W and W' makes (12.4.1) straightforward.

Furthermore, by definition, and since $\ker^q \Xi \supseteq \wedge^q W$ (cf. (12.3)),

$$\Xi \subseteq \ker^p(\ker^q \Xi) \subseteq \ker^p(\wedge^q W),$$

so (12.4.2) follows from (12.4.1).

Finally, assume that $q = 1$. In order to prove (12.4.3), we may obviously assume that $\dim \ker^q \Xi \geq p$ and let $W_1, W_2 \subseteq \ker^q \Xi$ be two subspaces of dimension p. Then by (12.4.2) $\wedge^p W_1 = \wedge^p W_2$, implying that $W_1 = W_2$. This can only happen, if $\dim \ker^q \Xi \leq p$. □

NOTATION 12.5. Let Z be a scheme, $\mathscr{E}_Z$ a locally free sheaf on Z, $p, q \in \mathbb{N}$, and $\mathscr{G} \subseteq \wedge^p \mathscr{E}_Z$ a subbundle. Observe that the natural map

$$\wedge^p \mathscr{E}_Z \otimes \wedge^q \mathscr{E}_Z \to \wedge^{p+q} \mathscr{E}_Z,$$

combined with the embedding $\mathscr{G} \hookrightarrow \wedge^p \mathscr{E}_Z$ induces a morphism:

$$\wedge_{\mathscr{G}} : \wedge^q \mathscr{E}_Z \to \mathscr{G}^* \otimes \wedge^{p+q} \mathscr{E}_Z.$$

We will also consider the corresponding subsheaf

$$\ker(\wedge_{\mathscr{G}}) \subseteq \wedge^q \mathscr{E}_Z.$$

We have the following.

Corollary 12.6. *Let $\mathscr{F} \subseteq \mathscr{E}_Z$ be a subbundle of rank r and $\mathscr{L} \subseteq \wedge^r \mathscr{E}_Z$ a line subbundle.*

(12.6.1) If $\mathscr{L} = \wedge^r \mathscr{F}$, then $\ker(\wedge_{\mathscr{L}}) = \mathscr{F}$.

(12.6.2) If $\mathscr{L} \subseteq \wedge^r \mathscr{E}_Z$ is such that $\ker(\wedge_{\mathscr{L}}) \supseteq \mathscr{F}$, then $\mathscr{L} = \wedge^r \mathscr{F}$.

(12.6.3) For any $\mathscr{L} \subseteq \wedge^r \mathscr{E}_Z$ and $z \in Z$, the rank of $(\ker(\wedge_{\mathscr{L}}))_z \subseteq (\mathscr{E}_Z)_z$ is at most r.

Proof. Clearly, if $\mathscr{L} = \wedge^r \mathscr{F}$, then $\ker(\wedge_{\mathscr{L}}) \supseteq \mathscr{F}$, and then (12.6.1) follows directly from (12.4.1) by restricting to fibers and setting $p = 1$ and $q = r$.

If $\mathscr{L} \subseteq \wedge^r \mathscr{E}_Z$ is such that $\ker(\wedge_{\mathscr{L}}) \supseteq \mathscr{F}$, then $\mathscr{L}$ and $\wedge^r \mathscr{F}$ are subbundles of the fixed locally free sheaf $\wedge^r \mathscr{E}_Z$, hence we only need to prove the statement fiberwise. Therefore we obtain that $\mathscr{L} \subseteq \wedge^r \mathscr{F}$ by (12.4.2) setting $p = r$ and $q = 1$. However, these are both line *subbundles*, so the containment must be an equality.

Finally, (12.6.3) is a straightforward consequence of (12.4.3) setting $p = r$ and $q = 1$. □

Proof of 12.1. Let $\mathbb{P} := \mathbb{P}(\wedge^r \mathscr{E}^*)$ with tautological line bundle $\mathscr{O}_{\mathbb{P}}(1)$. The dual of the natural surjective morphism (cf. [Har77, II.7.11]), $\wedge^r \mathscr{E}^*_{\mathbb{P}} \twoheadrightarrow \mathscr{O}_{\mathbb{P}}(1)$ gives an embedding

$$\mathscr{O}_{\mathbb{P}}(-1) \hookrightarrow \wedge^r \mathscr{E}_{\mathbb{P}}.$$

Now consider the morphism $\mathsf{M} := \wedge_{\mathscr{O}_{\mathbb{P}}(-1)}$ defined in (12.5). That is, twist the above embedding by $\mathscr{E}_{\mathbb{P}} \otimes \mathscr{O}_{\mathbb{P}}(1)$ to obtain

$$\mathscr{E}_{\mathbb{P}} \to \wedge^r \mathscr{E}_{\mathbb{P}} \otimes \mathscr{E}_{\mathbb{P}} \otimes \mathscr{O}_{\mathbb{P}}(1),$$

which, combined with the exterior product morphism, $\wedge^r \mathscr{E}_{\mathbb{P}} \otimes \mathscr{E}_{\mathbb{P}} \to \wedge^{r+1} \mathscr{E}_{\mathbb{P}}$, yields the natural morphism,

$$\mathsf{M} = \wedge_{\mathscr{O}_{\mathbb{P}}(-1)} : \mathscr{E}_{\mathbb{P}} \to \wedge^{r+1} \mathscr{E}_{\mathbb{P}} \otimes \mathscr{O}_{\mathbb{P}}(1).$$

Locally, M is given by a matrix of regular functions. Let $\mathscr{I}_r \subseteq \mathscr{O}_{\mathbb{P}}$ be the ideal sheaf generated by the $r \times r$-subdeterminants of these matrices, $\mathbb{G}_r \subset \mathbb{P}$ the subscheme defined by $\mathscr{I}_r$ and $\mathscr{K}_r = \ker \mathsf{M}\big|_{\mathbb{G}_r}$. Note that $\mathbb{G}_r$ is closed in $\mathbb{P}$ and hence projective over S.

Observe that by (12.6.3), the rank of $(\ker \mathsf{M})_x$ is at most r for any $x \in \mathbb{P}$. Then by construction $\mathscr{K}_r$ is a locally free sheaf of rank r. In fact, it follows from the construction that for any morphism $\zeta : Z \to \mathbb{P}$, $\ker(\zeta^* \mathsf{M}) = \zeta^*(\ker \mathsf{M})$ is a locally free sheaf of rank r if and only if ζ factors through $\mathbb{G}_r$.

Now let $Z \in \operatorname{Ob} \mathsf{Sch}_S$ and $\mathscr{F} \subseteq \mathscr{E}_Z$ be a rank r subbundle. Then $\mathscr{L} := \wedge^r \mathscr{F}$ is a subbundle of $\wedge^r \mathscr{E}_Z$, or equivalently there exists a surjective morphism $\wedge^r \mathscr{E}^*_Z \twoheadrightarrow \mathscr{L}^{-1}$. This induces an S-morphism $\zeta : Z \to \mathbb{P}$ such that $\mathscr{L}^{-1} = \zeta^* \mathscr{O}_{\mathbb{P}}(1)$. Therefore $\zeta^* \mathsf{M} = \wedge_{\mathscr{L}}$ and then it follows from (12.6.1) that $\mathscr{F} = \zeta^* \mathscr{K}_r$ and hence $(\mathbb{G}_r, \mathscr{K}_r)$ represents $\mathscr{G}rass_S(r, \mathscr{E})$. □

12.B The Hilbert functor

Let $g : Y \to Z$ be a projective morphism, $\mathscr{L}$ a g-ample line bundle on Y and $\mathscr{F}$ a coherent g-flat sheaf on Y. Then $\mathscr{R}^i g_*(\mathscr{F} \otimes \mathscr{L}^{\otimes m}) = 0$ for $i > 0$ and $m \gg 0$ [Har77, III.8.8(c)], so one has that

$$\operatorname{rk}(g_*(\mathscr{F} \otimes \mathscr{L}^{\otimes m}))_z = h^0(Y_z, \mathscr{F}_z \otimes \mathscr{L}_z^{\otimes m}) = \chi(Y_z, \mathscr{F}_z \otimes \mathscr{L}_z^{\otimes m}) \quad \text{for } z \in Z.$$

By the Riemann-Roch theorem the latter is a polynomial and since $\mathscr{F}$ is g-flat, this polynomial is independent of $z \in Z$ [Har77, III.9.9]. In other words, there exists a polynomial $h_{Y/Z,\mathscr{F},\mathscr{L}}$ such that

$$h_{Y/Z,\mathscr{F},\mathscr{L}}(m) = \operatorname{rk} g_*(\mathscr{F} \otimes \mathscr{L}^{\otimes m}) \qquad \text{for } m \gg 0.$$

We will call this the **Hilbert polynomial** of g with respect to $\mathscr{F}$ and $\mathscr{L}$. If there is no danger of confusion, then we will use the notation $h_{\mathscr{L}} := h_{Y/Z,\mathscr{O}_Y,\mathscr{L}}$ and will call $h_{\mathscr{L}}$ the Hilbert polynomial of $\mathscr{L}$.

It is important that we are considering projective morphisms and not simply morphisms with projective fibers. These two notions are different as shown by the following exercise.

EXERCISE 12.7. Give an example of a proper morphism $g : Y \to T$ such that Y_b is a projective variety for all $b \in B$, but the morphism g is not projective.

Let S be a scheme and $X \in \operatorname{Ob}\mathsf{Sch}_S$. We define the **Hilbert functor**,

$$\mathscr{H}\!\mathit{ilb}(X/S) : \mathsf{Sch}_S \to \mathsf{Sets}$$

as follows. For any $Z \in \operatorname{Ob}\mathsf{Sch}_S$,

$$\begin{aligned}\mathscr{H}\!\mathit{ilb}(X/S)(Z) :=& \{V \mid V \subseteq X \times_S Z \text{ flat and proper subscheme over } Z\} \\ \simeq& \{\mathscr{F} \mid \mathscr{F} \simeq \mathscr{O}_{X\times_S Z}/\mathscr{J} \text{ flat with proper support over } Z\},\end{aligned}$$

and for any $\phi \in \operatorname{Hom}_{\mathsf{Sch}_S}(Z, Y)$,

$$\begin{aligned}\mathscr{H}\!\mathit{ilb}(X/S)(\phi) : \mathscr{H}\!\mathit{ilb}(X/S)(Y) &\to \mathscr{H}\!\mathit{ilb}(X/S)(Z) \\ V &\mapsto V \times_Y Z \subseteq (X \times_S Y) \times_Y Z \simeq X \times_S Z.\end{aligned}$$

If $\mathscr{L}$ is a relatively ample line bundle on X/S and $p \in \mathbb{Q}[z]$, then we define

$$\mathscr{H}\!\mathit{ilb}_p(X/S)(Z) := \{\mathscr{F} \in \mathscr{H}\!\mathit{ilb}(X/S)(Z) \mid h_{X_Z/Z,\mathscr{F},\mathscr{L}_Z} = p\}.$$

Notice that if Z is connected, then

$$\mathscr{H}\!\mathit{ilb}(X/S)(Z) = \bigcup_p \mathscr{H}\!\mathit{ilb}_p(X/S)(Z).$$

Theorem 12.8 [Gro62, Gro95] [Kol96, I.1.4]. *Let X/S be a projective scheme, $\mathscr{L}$ a relatively ample line bundle on X/S and p a polynomial. Then the functor $\mathscr{H}ilb_p(X/S)$ is represented by a projective S-scheme* $\mathrm{Hilb}_p(X/S)$*, called the* **Hilbert scheme** *of X/S* **with respect to** p.

EXERCISE 12.9. Let X/S be a projective scheme. Prove that $\mathrm{Hilb}_1(X/S) \simeq X/S$.

REMARK 12.10. Similarly to (11.10.2-3), one observes that by the definition of representability, $\mathrm{id}_{\mathrm{Hilb}_p(X/S)} \in \mathrm{Hom}_{\mathrm{Sch}_S}(\mathrm{Hilb}_p(X/S), \mathrm{Hilb}_p(X/S))$ corresponds to a **universal object**, or **universal family**, $\mathrm{Univ}_p(X/S) \in \mathscr{H}ilb_p(X/S)(\mathrm{Hilb}_p(X/S))$. It follows from the definition of $\mathscr{H}ilb_p(X/S)$, that $\mathrm{Univ}_p(X/S) \subseteq X \times_S \mathrm{Hilb}_p(X/S)$ is flat and proper over $\mathrm{Hilb}_p(X/S)$ with Hilbert polynomial p.

DEFINITION 12.11. We define the **Hilbert scheme** of X/S as follows:

$$\mathrm{Hilb}(X/S) := \coprod_p \mathrm{Hilb}_p(X/S).$$

Sketch of the proof of 12.8. The idea of the construction of $\mathrm{Hilb}_p(X/S)$ is as follows.

Let $X \hookrightarrow \mathbb{P}$ be an embedding into a projective space over S such that a fixed power of $\mathscr{L}$ is the pull-back of $\mathscr{O}_{\mathbb{P}}(1)$. Further let $\mathscr{J} \subseteq \mathscr{O}_{\mathbb{P}}$ be the ideal sheaf of $X \subseteq \mathbb{P}$ and $m \in \mathbb{N}$ such that $\mathscr{J} \otimes \mathscr{O}_{\mathbb{P}}(m)$ is generated by global sections. Then the vector subspace

$$H^0(\mathbb{P}, \mathscr{J} \otimes \mathscr{O}_{\mathbb{P}}(m)) \subseteq H^0(\mathbb{P}, \mathscr{O}_{\mathbb{P}}(m))$$

determines $\mathscr{J} \otimes \mathscr{O}_{\mathbb{P}}(m) \subset \mathscr{O}_{\mathbb{P}}(m)$. By Mumford regularity [Mum66, Lecture 14] one may choose an m that works for all X with the same Hilbert polynomial.

Therefore we obtain a naturally defined subset of $\mathrm{Grass}_S(p(m), H^0(\mathbb{P}, \mathscr{O}_{\mathbb{P}}(m)))$ whose points are in one-to-one relationship with subschemes of $\mathbb{P}$ with Hilbert polynomial p. It can be proved that this subset is algebraic and has a natural subscheme structure cf. [Mum66, Lecture 8].

To finish the construction one proves that for $X = \mathbb{P}$ the subset of the Grassmanian with the obtained subscheme structure represents $\mathscr{H}ilb_p(\mathbb{P}/S)$. Then for arbitrary X, one argues that $\mathscr{H}ilb_p(X/S)$ is naturally a subfunctor of $\mathscr{H}ilb_p(\mathbb{P}/S)$ and thus $\mathrm{Hilb}_p(X/S)$ is constructed as a subscheme of $\mathrm{Hilb}_p(\mathbb{P}/S)$.

For more details the reader should consult [Kol96, I.1]. □

For more on Hilbert schemes see [Mum66, Kol96, EH00].

CHAPTER 13

The construction of the moduli space

13.A Boundedness

There are several properties a moduli functor needs to satisfy in order for it to admit a (coarse) moduli space cf. (1.A.8). We will discuss some of these in more detail. The first one is **boundedness**.

DEFINITION 13.1. Let $\mathcal{F}$ be a subfunctor of $\mathcal{MP}$. Then we say that $\mathcal{F}$ is **bounded** if there exists a scheme of finite type T and a family $(\pi : U \to T, \mathcal{L}) \in \mathcal{MP}(T)$ with the following property:

For any $(\sigma : X \to B, \mathcal{N}) \in \mathcal{F}(B)$ there exists an étale cover $\cup B_i \to B$ and finite type morphisms $\nu_i : B_i \to T$ such that for all i,

$$(\sigma : X \to B, \mathcal{N})_{B_i} \simeq \nu_i^*(\pi : U \to T, \mathcal{L}).$$

In this case we say that $(\pi : U \to T, \mathcal{L})$ is a **bounding family** for $\mathcal{F}$.

If in addition $(\pi : U \to T, \mathcal{L}) \in \mathcal{F}(T)$, then $(\pi : U \to T, \mathcal{L})$ is called a **locally versal family** for $\mathcal{F}$.

REMARK 13.2. When using canonical polarizations, then one may restrict to open covers in the definition. See [Vie95, 1.15] and [Kol94].

The first major general theorem about boundedness was Matsusaka's big theorem. Here we only cite a special case. For the more general statement please refer to the original article.

Theorem 13.3 [Mat72]. *Fix a polynomial $h \in \mathbb{Q}[t]$. Then $\mathcal{M}_h^{\text{smooth}}$ is bounded.*

In fact, in order to prove boundedness of $\mathcal{M}_h^{\text{smooth}}$, it is enough to prove the following:

Theorem 13.4. *Fix $h \in \mathbb{Q}[t]$. Then there exists an integer $m > 1$ such that $\omega_X^{\otimes m}$ is very ample for all $X \in \mathcal{M}_h^{\mathrm{smooth}}(\operatorname{Spec} k)$.*

DEFINITION 13.5. Let the smallest integer m satisfying the condition in (13.4) be denoted by $m(h)$.

Assume that (13.4) holds for m. Then by the Kodaira vanishing theorem (3.37) $\omega_X^{\otimes m}$ does not have higher cohomology for any $m \geq 2$ and $X \in \mathcal{M}_h^{\mathrm{smooth}}(\operatorname{Spec} k)$, so

$$h^0(X, \omega_X^{\otimes m}) = \chi(X, \omega_X^{\otimes m}) = h(m). \tag{13.5.1}$$

Choose $m = m(h)$ as defined in (13.5), that is, m is the smallest integer satisfying the condition in (13.4). Let $N = h(m) - 1$, the dimension of the projective space $\mathbb{P}(H^0(X, \omega_X^{\otimes m}))$ Then for all $X \in \mathcal{M}_h^{\mathrm{smooth}}(\operatorname{Spec} k)$ the m-th pluricanonical map

$$\phi_{\omega_X^{\otimes m}} : X \hookrightarrow \mathbb{P}^N$$

is an embedding. Next define the polynomial h' by $h'(x) := h(mx)$ and consider $T = \operatorname{Hilb}_{h'}(\mathbb{P}^N/k)$, $U = \operatorname{Univ}_{h'}(\mathbb{P}^N/k)$ and the two projections $\pi_1 : \mathbb{P}^N \times \operatorname{Hilb}_{h'}(\mathbb{P}^N/k) \to \mathbb{P}^N$ and $\pi_2 : \mathbb{P}^N \times \operatorname{Hilb}_{h'}(\mathbb{P}^N/k) \to \operatorname{Hilb}_{h'}(\mathbb{P}^N/k)$. Let $\pi = \pi_2|_U : U \to T$ and $\mathcal{L} = \pi_1^* \mathcal{O}_{\mathbb{P}^N/k}(1)|_U$. Then $(\pi : U \to T, \mathcal{L})$ gives a bounding family for $\mathcal{M}_h^{\mathrm{smooth}}$. Therefore (13.4) implies (13.3).

We will see later that it is necessary to allow singular objects in our moduli functors. This will lead to many difficulties, amongst them the fact that Matsusaka's big theorem, as stated, will not be strong enough for our purposes.

Recently the following more general boundedness statement was obtained through work of Tsuji, Hacon-McKernan and Takayama.

Theorem 13.6. *Fix $h \in \mathbb{Q}[t]$. Then there exists an integer $m > 0$ such that if X is a canonically polarized variety (i.e., ω_X is ample) with only canonical singularities and Hilbert polynomial h, then $\omega_X^{\otimes m}$ is very ample.*

This result is a consequence of the following weaker statement combined with the existence of canonical models.

Theorem 13.7. *Let $n > 0$ be any positive integer, then there exists an integer $r_n > 0$ and a real number $V_n > 0$ such that if X is a smooth complex projective variety of general type and dimension n, then:*

(13.7.1) $\operatorname{Vol}(K_X) > V_n$.

(13.7.2) For any $r \geq r_n$ the rational map $\phi_{|rK_X|} : X \dashrightarrow \mathbb{P}(H^0(\mathcal{O}_X(rK_X)))$ is birational.

Recall that by definition, the **volume** of K_X is given by

$$\operatorname{Vol}(K_X) = \text{lim.sup.}_{m\to\infty} \frac{h^0(\mathcal{O}_X(mK_X))}{m^n/n!}$$

where $n = \dim X$. It is known that the above lim.sup. is in fact a limit and that X is of general type if and only if $\operatorname{Vol}(K_X) > 0$.

Notice that the first statement in the above theorem is immediate from the second statement. In fact, we have that if $\phi_{|r_n K_X|}$ is birational, then

$$\mathrm{Vol}(r_n K_X) \geq \deg\left(\overline{\phi_{|r_n K_X|}(X)} \subset \mathbb{P}H^0(\mathcal{O}_X(r_n K_X))\right) \geq 2$$

and so $\mathrm{Vol}(K_X) > (1/r_n)^n$.

By the same line of argument as above (cf. (13.4)), we have the following immediate corollary of (13.7).

Corollary 13.8. *Let $n > 0$ be any positive integer and $M > 0$ be any positive real number. Then the set of complex projective n-dimensional varieties X such that $0 < \mathrm{Vol}(K_X) \leq M$ is birationally bounded in the sense that there exists a projective morphism of normal quasi-projective varieties (in particular of finite type over $\mathbb{C}$) $f : \mathcal{Z} \to B$ such that for any X as above, there exists a closed point $b \in B$ such that the fiber $\mathcal{Z}_b$ is birational to X.*

We also have the following more detailed corollary.

Corollary 13.9. *Let $n > 0$ be any positive integer and $M > 0$ be any positive real number, then the set of complex projective projective varieties X such that*

(13.9.1) X is a canonically polarized variety with only canonical singularities;

(13.9.2) $\dim X = n$; and

(13.9.3) $\mathrm{Vol}(K_X) \leq M$,

has a parameter space; i.e., there exists a projective morphism of normal quasi-projective varieties $f : \mathcal{X} \to B$ such that for any X as above, there is a closed point $b \in B$ such that the fiber $\mathcal{X}_b$ is isomorphic to X and such that any fiber $\mathcal{X}_b$ is a canonically polarized variety with only canonical singularities.

Proof. Let $f : \mathcal{Z} \to B$ be the family given by (13.8). We may assume that B is the disjoint union of its irreducible components. We may replace B by the closure in B of the set of points

$$\{b \in B | \mathcal{Z}_b \text{ is a variety of general type}\} \subset B.$$

Let η be the generic point of a component of B and let $\mathcal{Y}_\eta \to \mathcal{Z}_\eta$ be a projective resolution of the singularities of the fiber over η. By (5.60), $R(K_{\mathcal{Y}_\eta})$ is finitely generated. Let N_η be an integer such that $R(N_\eta K_{\mathcal{Y}_\eta})$ is generated in degree 1. The above resolution extends to an open neighborhood $U = U_\eta$ of η in B and we may assume that $R(N_\eta K_{Y_U})$ is generated in degree 1. Therefore, by Noetherian induction, replacing B by a disjoint union of locally closed subsets and replacing $\mathcal{Z}$ by a desingularization, we may assume that $f : \mathcal{Z} \to B$ is a smooth projective morphism and that there exists an integer N such that for any closed point $b \in B$, $R(NK_{\mathcal{Z}_b})$ is generated in degree 1 and

$$\phi_{NK_{\mathcal{Z}_b}} : \mathcal{Z}_b \dashrightarrow \mathcal{Z}_b^{\mathrm{can}} \cong X_b$$

defines a birational map to the canonical model. The statement now follows by letting $\mathcal{X} = \mathrm{Proj}_B R(K_{\mathcal{Z}}/B)$. □

Proof of (13.6). Note that $\dim X$ and $\mathrm{Vol}(K_X)$ are determined by the degree and the leading coefficient of the polynomial h. From (13.9) and its proof it follows that there is an integer $N > 0$ such that NK_X is Cartier. By (5.3), there exists an integer m (depending only on n) such that mNK_X is very ample. □

REMARK 13.10. Notice that this is enough to deal with canonical models of smooth projective varieties of general type, in particular to construct a coarse moduli space for $\mathcal{M}_h^{\text{smooth}}$. However, unfortunately, this does not lead to a compact moduli space. In order to construct a moduli space for all stable varieties, one needs a similar boundedness statement for a larger class of singularities. We will not discuss this here, the reader may consult [Ale94], [Ale02], [AM04] and [Kol].

The rest of the present section is devoted to proving (13.7), but first we will need to recall several important results.

Lemma 13.11. *Let X be a variety of general type and $V_x \subset X$ a subvariety containing a very general point $x \in X$. Then V_x is of general type.*

Proof. Suppose that V_x is not of general type. Since x is general, we may assume that the V_x are a dominating family of subvarieties of X cf. (2.G). Therefore, there is a flat projective morphism $p : \mathcal{V} \to B$ of normal quasi-projective varieties and a dominating morphism $f : \mathcal{V} \to X$ such that there is a dense subset $B^0 \subset B$ (which we may assume is not contained in a countable union of closed subsets) such that for any closed point $b \in B^0$, the fiber $\mathcal{V}_b$ is isomorphic to a V_x and hence is not of general type. Cutting down by hyperplanes, compactifying, and resolving the indeterminacies, we may assume that f is a generically finite morphism of smooth projective varieties (p is no longer flat).

We have that $K_{\mathcal{V}} = f^*K_X + R$ where R is an effective divisor. Therefore, $\mathcal{V}$ is of general type. By the Easy Addition Formula, it then follows that the general fibers of p are of general type. This contradicts our assumptions. □

Lemma 13.12. *Let (X, D) be a quasi-projective $\mathbb{Q}$-factorial lc pair, V a minimal center of non-klt singularities not contained in the non-klt singularities of $(X, 0)$. Let H be an ample divisor on X and assume that for a general point $v \in V$, there exists a divisor $\bar{H}_v \sim_{\mathbb{Q}} H|_V$ such that $\mathrm{mult}_v(\bar{H}_v) > \dim V$, then there exists a $\mathbb{Q}$-divisor $G \sim_{\mathbb{Q}} \alpha D + \beta H$ where $0 \leq \alpha < 1$, $0 < \beta < 1$, (X, G) is lc on a neighborhood of v and v is contained in a unique non-klt center V' of (X, G) with $\dim V' < \dim V$.*

Proof. Note that on a neighborhood of v, the pair $(X, (1-\varepsilon)D)$ is klt for any $0 < \varepsilon \leq 1$. Let $f : Y \to X$ be a log resolution of (X, D) and write $K_Y + \Gamma = f^*(K_X + D) + E$ as in (3.21.1). By assumption, there is a component F of Γ of coefficient 1 such that $f(F) = V$. Since H is ample, by Serre vanishing, for general $v \in V$, we may find a divisor $H_v \sim_{\mathbb{Q}} H$ such that $H_v|_V = \bar{H}_v$ and $(X, D + H_v)$ is lc on the complement of V. Let F_v be the fiber of F over v. Since $v \in V$ is general, we may assume that $F \to V$ is smooth over a neighborhood of v and so $\mathrm{mult}_{F_v}(f^*H_v|_F) > \dim V$. In particular $(F, f^*H_v|_F)$ is not lc along F_v and hence $(F, (\Gamma - F + f^*H_v)|_F)$ is also not lc. Let $f' = f \circ \mu : Y' \to Y \to X$ be a log resolution of $(X, D + H_v)$. We write $K_{Y'} + \Gamma' = f'^*(K_X + \Delta) + E'$ as in (3.21.1) and $F' = \mu_*^{-1}F$. As $(F, (\Gamma - F + f^*H_v)|_F)$

is not lc, neither is $(F', (\Gamma' - F' + f'^* H_v)|_{F'})$. Thus there is a component $\Gamma'_0 \neq F'$ of multiplicity greater than 1 in Γ' whose intersection with F' dominates v. It follows that (in a neighborhood of v), for some $0 \leq \alpha < 1$, the pair $(X, \alpha D + H_v)$ is klt over $X - V \cap \operatorname{Supp}(H_v)$ and $(X, \alpha D + H_v)$ is not lc over a center containing v. Therefore, there exists a number $0 \leq \beta < 1$ such that $(X, \alpha D + \beta H_v)$ is lc but not klt over a center containing v. Let $v \in V' \subset V$ be any minimal non klt center containing v. By tie braking, cf. [Kol07a, 8.7], for any $\varepsilon > 0$, there exists a divisor $D' \sim \tau(\alpha D + \beta H_v + \varepsilon H)$ where $0 < \tau < 1$ and (X, D') has a unique non-klt place over V'. The lemma now follows. □

Theorem 13.13 Kawamata's subadjunction theorem. *Let $V \subset X$ be a minimal center of non-klt singularities for a pair (X, D) which is lc at a general point of V. Let $\mu : \tilde{V} \to V$ be the normalization and let G be an ample $\mathbb{Q}$-Cartier divisor on X. Then for any rational number $\varepsilon > 0$, there is an effective $\mathbb{Q}$-divisor Δ_ε such that*

$$\mu^*(K_X + D + \varepsilon G)|_V = K_{\tilde{V}} + \Delta_\varepsilon.$$

If moreover, V is a minimal non-klt center at $v \in V$ and (X, D) is lc at $v \in V$, then V is normal and (V, Δ_ε) is klt on a neighborhood of $v \in V$.

Proof. See [Kol07a, 8.6]. □

Proof of (13.7). We proceed by induction on $n = \dim X$. Note that when $n = 1$, $\operatorname{Vol}(K_X) \geq 2$ and $\phi_{|rK_X|}$ defines an embedding for all $r \geq 3$.

Step 1. *It suffices to show that there exist positive constants $A, B > 0$ such that $\phi_{|rK_X|}$ is birational for all*

$$r \geq \frac{A}{(\operatorname{Vol}(K_X))^{1/n}} + B.$$

If $\operatorname{Vol}(K_X) \geq 1$, then $\phi_{|rK_X|}$ is birational for all $r \geq A + B$. If $\operatorname{Vol}(K_X) < 1$, then let $r_0 = \left\lceil \frac{A}{(\operatorname{Vol}(K_X))^{1/n}} + B \right\rceil$. Then $\phi_{|r_0 K_X|}$ is birational and

$$\operatorname{Vol}(r_0 K_X) = r_0^n \operatorname{Vol}(K_X) < \left(\frac{A}{(\operatorname{Vol}(K_X))^{1/n}} + B + 1 \right)^n \operatorname{Vol}(K_X) < (A + B + 1)^n.$$

It follows that the closure of $\phi_{|r_0 K_X|}(X)$ has bounded degree and hence belongs to a bounded family. We may therefore assume that there is a projective morphism of quasi-projective varieties $f : \mathscr{X} \to B$ such that if X is any smooth complex projective variety of dimension n and $0 < \operatorname{Vol}(K_X) < 1$, then X is birational to $\mathscr{X}_b$, the fiber of f over a closed point $b \in B$. By an argument similar to the one in the proof of (13.9), we may assume that f is smooth. If $\phi_{|rK_{\mathscr{X}_b}|}$ is birational for some closed point $b \in B$, then we may assume that the same is true for any closed point b' in a neighborhood of b. By Noetherian induction, there exists a constant r_n such that $\phi_{|rK_X|}$ is birational for any $r \geq r_n$ and any X smooth complex projective variety of dimension n with $0 < \operatorname{Vol}(K_X) < 1$.

Step 2. *It suffices to show that there exist positive constants $A, B > 0$ such that for any very general point $x \in X$, there is a $\mathbb{Q}$-divisor $D_x \sim_{\mathbb{Q}} \lambda K_X$ where*

- *x is an isolated point of the co-support of $\mathcal{J}(D_x)$;*
- $\lambda < \frac{A}{(\mathrm{Vol}(K_X))^{1/n}} + B$.

Let $m \gg 0$ be an integer such that $|mK_X|$ defines a birational map and pick $mG \in |mK_X|$ a very general divisor. Let $(x, y) \in X \times X$ be a very general point. In particular x and y are very general points of X. Then x is not contained in the support of $D_y + G$ and y is not contained in the support of $D_x + G$. Let $t = \left\lceil \frac{A}{(\mathrm{Vol}(K_X))^{1/n}} + B \right\rceil$, then x and y are isolated points in the co-support of $\mathcal{J}(D_x + D_y + (t - 2\lambda)G)$. Let $\mu : Y \to X$ be a log resolution of $(X, D_x + D_y + G)$, then we may write $f^*G = F + M$ where $mM = \mathrm{Mob}(\mu^* mK_X)$ is free and hence nef and big. Note that if $\lfloor (t - 2\lambda)M \rfloor = 0$ then

$$t\mu^* K_X - \lfloor \mu^* D_x + \mu^* D_y + (t - 2\lambda)\mu^* G \rfloor \equiv \{\mu^* D_x + \mu^* D_y + (t - 2\lambda)F\} + (t - 2\lambda)M.$$

By Kawamata-Viehweg vanishing, it follows that

$$\mathcal{R}^1 \mu_* \mathcal{O}_Y(K_Y + t\mu^* K_X - \lfloor \mu^* D_x + \mu^* D_y + (t - 2\lambda)\mu^* G \rfloor) = 0$$

and hence (again by Kawamata-Viehweg vanishing and an easy spectral sequence argument) that

$$\begin{aligned} &H^1(X, \mathcal{O}_X((t+1)K_X) \otimes \mathcal{J}(D_x + D_y + (t - 2\lambda)G)) \\ &= H^1(Y, \mathcal{O}_Y(K_Y + t\mu^* K_X - \lfloor \mu^* D_x + \mu^* D_y + (t - 2\lambda)\mu^* G \rfloor)) = 0. \end{aligned}$$

But then the homomorphism

$$H^0(\mathcal{O}_X((t+1)K_X)) \to H^0(\mathcal{O}_X((t+1)K_X) \otimes \mathcal{O}_X / \mathcal{J}(D_x + D_y + (t - 2\lambda)G))$$

is surjective. In particular $H^0(\mathcal{O}_X((t+1)K_X)) \geq 2$ and we have a rational map $\phi : X \dashrightarrow \mathbb{P}^1$. Replacing X by an appropriate smooth birational model, we may assume that ϕ is a morphism. By induction, if X_p is the fiber over a general point $p \in \mathbb{P}^1$, then $|rK_{X_p}|$ induces a birational map for all $r \geq r_{n-1}$. By [Kol86], we may choose

$$r_n = (2 \left\lceil \frac{A}{(\mathrm{Vol}(K_X))^{1/n}} + B \right\rceil + 1)(2r_{n-1} + 2) + r_{n-1}.$$

Step 3. *Constructing D_x.*

Replacing X by its canonical model, we may assume that K_X is ample and X has canonical singularities. Since $h^0(\mathcal{O}_X(mK_X)) = vm^n/n! + O(m^{n-1})$ where $v = \mathrm{Vol}(K_X)$, and vanishing to order k at a very general point $x \in X$ imposes at most $k^n/n!$ conditions, for any constant $0 < v' < v$, there exists an integer $m > 0$ such that for any very general point $x \in X$, we may find a divisor $D_x \in |mK_X|$ with $\mathrm{mult}_x(D_x) > m(v')^{1/n}$. Let

$$\tau = \sup\{t > 0 | (X, tD_x) \text{ is klt near } x\}.$$

Thus $\tau \leq n/(m(v')^{1/n})$ and so $\tau D_x \sim_{\mathbb{Q}} \eta K_X$ where $\eta \leq n/(v')^{1/n}$. Let V_x be a minimal non-klt center of $(X, \tau D_x)$ at x and let $\mu : \tilde{V}_x \to V_x$ be the normalization. If $\dim V_x = 0$

we are done, so we may assume that $n > n' = \dim V_x > 0$. Since $x \in X$ is general, by (13.11), we have that V_x is of general type.

Let $W_x \to \tilde{V}_x$ be a resolution. By induction on the dimension, we may assume that

$$\mathrm{Vol}(K_{W_x}) \geq V_{n'}.$$

Therefore, for general $w \in W_x$, there is a divisor $G_w \sim \eta' K_{W_x}$ such that $\mathrm{mult}_w(G_w) > \dim W_x$ and $\eta' < n/(V_{n'})^{1/n'}$. This pushes forward to a divisor $\bar{G}_w \sim_{\mathbb{Q}} \eta' K_{\tilde{V}_x}$ with the same properties. By (13.13), for any rational number $\varepsilon > 0$ and any general point $w \in V_x$, we may find a divisor $\bar{H}_w \sim_{\mathbb{Q}} \eta'(K_X + \tau D_x + \varepsilon K_X)|_{V_x}$ such that $\mathrm{mult}_w(\bar{H}_w) > \dim W_x$. By (13.12), for any general point $v \in V$, there exists a $\mathbb{Q}$-divisor G on X such that

$$G \sim_{\mathbb{Q}} \lambda K_X \qquad \text{where } \lambda < \eta + (1 + \eta + \varepsilon)\eta'$$

and (X, G) is lc on a neighborhood of $v \in X$ and has a unique a non-klt center V' containing v with $\dim V' < \dim V$.

Repeating this process at most $\dim V_x$ times, one sees that we may assume that there exist constants $\alpha, \beta > 0$ such that w is the unique non-klt center of (X, D_w) on a neighborhood of a general point $w \in X$ and that $D_w \sim_{\mathbb{Q}} \lambda K_X$ where $\lambda < \alpha\eta + \beta$. The claim now follows easily as the first summand is of the form $O(1/(\mathrm{Vol}(K_X))^{1/n})$. □

13.B Constructing the moduli space

The success of using the Hilbert scheme in order to obtain boundedness might make one believe that the Hilbert scheme itself could work as a moduli space. However, unfortunately this is not the case as the points of $\mathrm{Hilb}_h(\mathbb{P}^N/k)(k)$ also parameterize subschemes that are not in the moduli functor $\mathcal{M}_h^{\mathrm{smooth}}$. For example, they may be horribly singular and the polarizing line bundle may not even be related to the canonical sheaf.

The next natural idea is to take the locus of Hilbert points that corresponds to those subvarieties of $\mathbb{P}^N$ that are in $\mathcal{M}_h^{\mathrm{smooth}}(\mathrm{Spec}\, k)$, i.e., smooth with canonical polarization. This is a much better guess, but still not perfect. There are two fundamental problems. First, it is not at all clear that this locus is a subscheme of $\mathrm{Hilb}_h(\mathbb{P}^N/k)$, or even if its support is a subscheme, then whether there is a natural scheme structure that is compatible with the functor $\mathcal{M}_h^{\mathrm{smooth}}$. This actually turns out to be a difficult technical problem referred to as **local closedness** and we will discuss it in more detail below. The second problem is that a single object of $\mathcal{M}_h^{\mathrm{smooth}}(\mathrm{Spec}\, k)$ will appear several times in $\mathrm{Hilb}_h(\mathbb{P}^N/k)$; any subscheme of $\mathbb{P}^N$ appears as a potentially different subscheme after acting upon it by an element of $\mathrm{Aut}(\mathbb{P}^N)$, but in the moduli functor we only want a single copy of each isomorphism class.

The way to proceed is "obvious". Assume that we can solve the local closedness problem and indeed we can find a subscheme that consists of exactly the points that belong to $\mathcal{M}_h^{\mathrm{smooth}}(\mathrm{Spec}\, k)$. (Actually we will need to worry about more than that.) Then we get a natural action of $\mathrm{Aut}(\mathbb{P}^N)$ on this subscheme and taking the quotient by $\mathrm{Aut}(\mathbb{P}^N)$ will yield our desired moduli space. It should be mentioned that it is not entirely obvious how one should go about taking this quotient, but fortunately it has been figured out and explained in [Vie91, Kol97a, KeM97].

As we mentioned before, $\mathcal{M}_h^{\text{smooth}}$ does not lead to a compact moduli space. As in the case of curves, this is settled by extending the moduli functor to allow **stable** varieties (cf. 13.26) instead of just smooth ones. We will only briefly touch on this extension in the following sections, but we will at least make some remarks indicating some of the issues and difficulties related to working with the moduli functor of stable varieties.

13.C Local closedness

We have already observed that in order to carry out the the plan laid out in 13.B we need to identify the set of Hilbert points corresponding to the moduli functor and find a natural scheme structure on this set. The technical condition that allows us to do this is the following.

DEFINITION 13.14. A subfunctor $\mathcal{F} \subseteq \mathcal{MP}$ is **locally closed** (resp. **open**, **closed**) if the following condition holds: For every $(f : X \to B, \mathcal{L}) \in \mathcal{MP}(B)$ there exists a locally closed (resp. open, closed) subscheme $\iota : B' \hookrightarrow B$ such that if $\tau : T \to B$ is any morphism then

$$(f_T : X_T \to T, \mathcal{L}_T) \in \mathcal{F}(T) \quad \Leftrightarrow \quad \tau \text{ factors through } \iota \qquad \begin{array}{ccc} T & \xrightarrow{\tau} & B \\ & \searrow^{\exists} & \nearrow_{\iota} \\ & B' & \end{array} .$$

OBSERVATION 13.15. There are two main ingredients of proving that $\mathcal{M}_h^{\text{smooth}}$ is locally closed. Let $m = m(h)$ as defined in (13.5). Suppose that $(f : X \to B, \mathcal{L}) \in \mathcal{MP}(B)$. Note that in the construction of the moduli space this $\mathcal{L}$ comes from $\mathcal{O}_{\mathbb{P}^N}(1)$ where $N = h(m) - 1$. Now one needs to prove that:

(13.15.1) the set $\{b \in B \mid X_b \in \mathcal{M}_h^{\text{smooth}}(\operatorname{Spec} k)\}$ is a locally closed subset of B, and

(13.15.2) the condition $\mathcal{L}\big|_{X_b} \simeq \omega_{X_b}^m$ is locally closed on B.

At this point these conditions are not too hard to satisfy. To prove (13.15.1) one observes that being projective is assumed, while being smooth is open. The canonical bundle, ω_{X_b} being ample is open, but this we actually do not even need as it will follow from (13.15.2). The requirement on the Hilbert polynomial will also follow from (13.15.2). In turn, (13.15.2) follows from the following lemma.

Lemma 13.16 [Vie95, 1.19]. *Let $f : X \to B$ be a flat projective morphism and $\mathcal{K}$ and $\mathcal{L}$ two line bundles on X. Assume that $h^0(X_b, \mathcal{O}_{X_b}) = 1$ for all $b \in B$. Then there exists a locally closed subscheme $\iota : B' \hookrightarrow B$ such that if $\tau : T \to B$ is any morphism then*

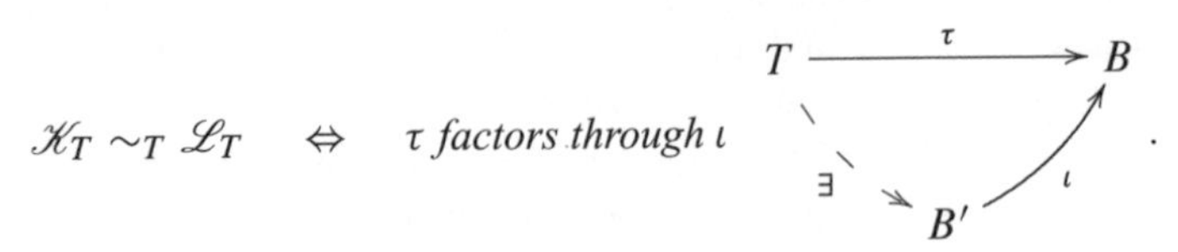

Proof. After replacing $\mathscr{L}$ by $\mathscr{L} \otimes \mathscr{K}^{-1}$ we may assume that $\mathscr{K} \simeq \mathscr{O}_X$. Observe that if $\mathscr{L}\big|_{X_b}$ is generated by a single section, then it gives an isomorphism

$$\mathscr{O}_{X_b} \xrightarrow{\simeq} \mathscr{L}\big|_{X_b}.$$

Consider

$$B''_{\text{red}} := \left\{ b \in B \mid h^0(X_b, \mathscr{L}\big|_{X_b}) \neq 0 \right\} = \operatorname{supp}(f_*\mathscr{L}).$$

This is closed by semi-continuity [Har77, III.12.8]. So far this is only a subset and we need to define a (natural) scheme structure on it. However, that is a local problem, so we may assume that B is affine. By cohomology-and-base-change [Mum70, §5] there exists a bounded complex of locally free sheaves

$$\mathscr{E}^0 \xrightarrow{\delta^0} \mathscr{E}^1 \xrightarrow{\delta^1} \cdots \xrightarrow{\delta^{n-1}} \mathscr{E}^n$$

such that for any morphism $\tau : T \to B$,

$$\mathscr{R}^i(f_T)_*\mathscr{L}_T \simeq h^i(\mathscr{E}_T^{\bullet}).$$

In particular,

$$(f_T)_*\mathscr{L}_T \simeq \ker[\delta_T^0 : \mathscr{E}_T^0 \to \mathscr{E}_T^1].$$

By definition, $B''_{\text{red}} = \operatorname{supp}\ker\delta^0$. Now define the ideal sheaf $\mathscr{I} \subseteq \mathscr{O}_B$ as follows:

- If $B''_{\text{red}} = B_{\text{red}}$ in a neighbourhood of a point $b \in B$, then let $\mathscr{I} = 0$ near b.
- Otherwise write $\mathscr{E}^i = \oplus^{r_i}\mathscr{O}_B$ near $b \in B$. Since we are not in the previous case, we must have $r_1 \geq r_0$. Now let $\mathscr{I}$ be generated by the $r_0 \times r_0$ minors of

$$\delta^0 : \bigoplus^{r_0} \mathscr{O}_B \to \bigoplus^{r_1} \mathscr{O}_B.$$

Let the scheme structure on B''_{red} be defined by this ideal sheaf, i.e., let B'' the scheme with support equal to B''_{red} and whose structure sheaf is $\mathscr{O}_{B''} := \mathscr{O}_B \big/ \mathscr{I}$.

Now if $\tau : T \to B$ is such that $\mathscr{L}_T \sim_T \mathscr{O}_T$, then $(f_T)_*\mathscr{L}_T$ is a line bundle on T and if $\ker\delta_T^0$ contains a line bundle, then the image of $\tau^*\mathscr{I}$ in $\mathscr{O}_T$ has to be zero. In other words, τ factors through $B'' \hookrightarrow B$.

In the final step we construct B' as an open subscheme of B''. By our previous observation we may assume that $B'' = B$, in particular, $f_*\mathscr{L} \neq 0$ on a dense open set. Let

$$B''' := \left\{ b \in B \mid h^0(X_b, \mathscr{L}\big|_{X_b}) > 1 \right\}.$$

Again, by semi-continuity, B''' is closed. Next let $B^\circ = B \setminus B'''$, the largest open (possibly empty) subscheme of B with $f_*\mathscr{L}\big|_{B^\circ}$ invertible and let $Z \subseteq X$ be the support of $\operatorname{coker}[f^*f_*\mathscr{L} \to \mathscr{L}]$. Finally let

$$B' := (B \setminus f(Z)) \cap B^\circ \subseteq B.$$

It is easy to check that this B' satisfies the required condition. □

Extending local closedness to the moduli functor of stable varieties is an extremely hard problem and has been open for a long time. It has been recently solved by Kollár in [Kol08a]. See also [HK04].

13.D Separatedness

Boundedness and local closedness allow us to identify a subscheme of an appropriate Hilbert scheme consisting of the Hilbert points of the schemes in our moduli problem. This subscheme has a group action induced by the automorphism group of the ambient projective space. This already allows the construction of the moduli space as an algebraic space by taking the quotient by this group action. However, in order to effectively use this moduli space we hope that it will satisfy certain basic properties. Perhaps the most basic one is separatedness.

DEFINITION 13.17. A subfunctor $\mathscr{F} \subseteq \mathscr{M\!P}$ is **separated** if the following condition holds. Let R be a DVR and $T = \operatorname{Spec} R$ with general point $t_g \hookrightarrow T$ and $(X_i \to T, \mathscr{L}_i) \in \mathscr{F}(T)$ two families for $i = 1, 2$. Then any isomorphism $\alpha_g : ((X_1)_{t_g}, (\mathscr{L}_1)_{t_g}) \to ((X_2)_{t_g}, (\mathscr{L}_2)_{t_g})$ over t_g extends to an isomorphism $\alpha : X_1 \to X_2$ over T.

Separatedness of a moduli functor is a non-trivial property. Without further restrictions it will not hold as shown by the following examples. We learned these examples from János Kollár.

EXAMPLE 13.18. Let $Z = \mathbb{P}^1 \times \mathbb{A}^1$ with coordinates $([x : y], t)$. Let the projections to the factors be $\pi_1 : Z \to \mathbb{P}^1$ and $\pi_2 : Z \to \mathbb{A}^1$. Further let $\mathscr{L} = \pi_1^* \mathscr{O}_{\mathbb{P}^1}(1)$, $R = k[t]_{(t)}$ (a DVR) and consider the base change by $T = \operatorname{Spec} R \to \mathbb{A}^1$. With the notation $f = (\pi_2)_T$, one has that $(f : Z_T \to T, \mathscr{L}_T) \in \mathscr{M\!P}(T)$. Now let $\alpha : Z_T \dashrightarrow Z_T$ be the map induced by $([x : y], t) \mapsto ([tx : y], t)$. This is an isomorphism over the general point of T, but is not even dominant over the special point.

REMARK 13.19. The main problem here comes from the fact that $\operatorname{Aut} \mathbb{P}^1$ is not discrete. The good news is that by a theorem of Matsusaka and Mumford [MM64] this problem can only occur if the fiber over the closed point is ruled.

EXAMPLE 13.20. Let Y be a smooth projective variety of dimension at least 2, $Z = Y \times \mathbb{A}^1$, $\pi : Z \to \mathbb{A}^1$ the projection to the second factor and $C_1, C_2 \subseteq Z$ two sections, i.e., curves in Z that are isomorphic to $\mathbb{A}^1$ via π and such that C_1 and C_2 intersect in a single point, P, transversally. Assume for simplicity that $\pi(P) = 0 \in \mathbb{A}^1$.

Let Z_1 be the variety obtained by first blowing up C_1 and then the strict transform of C_2. Similarly, let Z_2 be the variety obtained by first blowing up C_2 and then the strict transform of C_1.

Let $U = \{t \in \mathbb{A}^1 \mid t \neq 0\}$. Then $(Z_1)_U$ and $(Z_2)_U$ are naturally identified, but this identification between $(Z_1)_U$ and $(Z_2)_U$ does not extend over $t = 0 \in \mathbb{A}^1$. To see this, observe that the exceptional divisors on the fibers $(Z_i)_0$ are distinguishable. If $\dim Y = 2$ then the exceptional fiber corresponding to the first blow-up has self-intersection -2 while the other one has self-intersection -1. If $\dim Y = d > 2$ then the two exceptional divisors are not even isomorphic. The one coming from the second blow-up is isomorphic to $\mathbb{P}^{d-1}$

while the one coming from the first is the blow-up of that at a point. On the other hand, any extension of the identification between $(Z_1)_U$ and $(Z_2)_U$ would map the preimage of C_i to the preimage of C_i for $i = 1, 2$ and hence would have to identify the first exceptional divisor on $(Z_1)_0$ with the second one on $(Z_2)_0$. This is obviously impossible.

To make this example more interesting, assume that $\mathbb{A}^1$ admits an embedding into $\operatorname{Aut} Y$, i.e., Y admits a one-parameter group of automorphisms. Denote these automorphisms by α_t for $t \in \mathbb{A}^1$ and assume that $\alpha_0 = \mathrm{id}_Y$ and $(C_2)_t = \alpha_t((C_1)_t)$. In this case the automorphisms α_t induce an isomorphism between Z_1 and Z_2 including the fiber over $t = 0$. Observe that the restriction of this isomorphism is the identity on the fiber over $t = 0$, but different from the identity over any $t \in U$.

In this example Z_1 and Z_2 are isomorphic, but not all isomorphisms over U extend to an isomorphism over the entire $\mathbb{A}^1$.

EXAMPLE 13.21. This example is based on an example of Atiyah. Let $\iota : \mathbb{P}^1 \times \mathbb{P}^1 \hookrightarrow \mathbb{P}^n$ be an arbitrary embedding and $Y \subseteq \mathbb{P}^{n+1}$ the projectivized cone over $\iota(\mathbb{P}^1 \times \mathbb{P}^1) \subseteq \mathbb{P}^n$ with vertex P. Let $L \subseteq \mathbb{P}^{n+1}$ be a general linear subspace of codimension 2. Notice that this implies that $P \notin L$. Consider the projection from L to a line, $\mathbb{P}^{n+1} \setminus L \to \mathbb{P}^1$. After blowing up L this extends to a morphism $\pi_L : \mathrm{Bl}_L \mathbb{P}^{n+1} \to \mathbb{P}^1$. Let Z be the strict transform of Y on $\mathrm{Bl}_L \mathbb{P}^{n+1}$ and $\pi = \pi_L\big|_Z$. Then one has the following diagram:

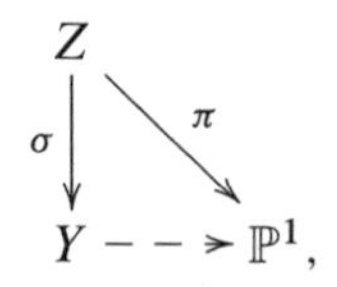

where π is flat projective with connected fibers and a smooth general fiber, and σ is the blowing-up of $L \cap Y \subseteq Y$, hence birational and an isomorphism near $P \in Y$. Let $\widetilde{P} = \sigma^{-1}(P)$.

Next let C_1 and C_2 be the images via ι of two general lines corresponding to the two different rulings of $\mathbb{P}^1 \times \mathbb{P}^1$ and S_1 and S_2 their respective preimages on Y. Note that by construction C_1 and C_2 are disjoint from L. For the rest of this example anywhere i appears, it is meant to apply for both $i = 1, 2$. Let $\widetilde{S}_i = \sigma^{-1} S_i \subseteq Z$ and $\sigma_i : Z_i = \mathrm{Bl}_{\widetilde{S}_i} Z \to Z$ the blow-up of Z along $\widetilde{S}_i$. Observe, that $\widetilde{S}_i \subseteq Z$ is a divisor and since Z is smooth away from $\widetilde{P}$, this implies that Z_i is isomorphic to Z away from $\widetilde{P}$, in particular $(Z_1)_{\mathbb{P}^1 \setminus \{Q\}} \simeq (Z_2)_{\mathbb{P}^1 \setminus \{Q\}}$ where $Q = \pi(\widetilde{P}) \in \mathbb{P}^1$. On the other hand, it is easy to check that $\sigma_i^{-1}(\widetilde{P}) \simeq \mathbb{P}^1$ is equal to the whole fiber of the blow-up $\mathrm{Bl}_{S_i} \mathbb{P}^{n+1} \to \mathbb{P}^{n+1}$. (Since $P \notin L$, this computation can be done on Y).

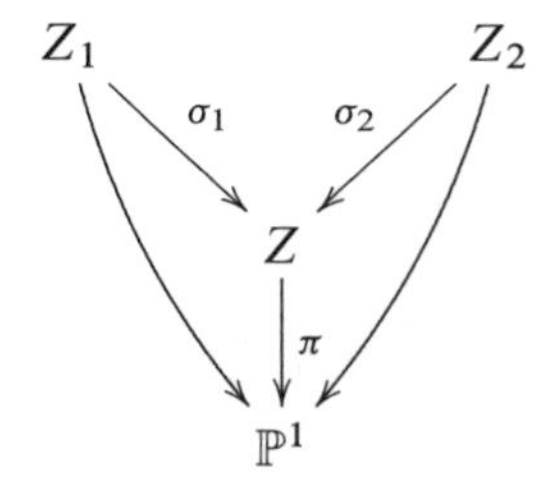

Next, we wish to determine the fiber $Z_Q = \pi^{-1}(Q)$. Let $L_P = \langle L, P \rangle$, the linear span of L and P. Observe that $L_P \simeq \mathbb{P}^n$ is a general hyperplane through P in $\mathbb{P}^{n+1}$. Hence $L_P \cap Y$ is the cone over a general hyperplane section of a smooth projective surface, i.e., over a smooth projective curve. We conclude that Z_Q is the blow-up of this cone at its intersection with L which consists of finitely many points that are disjoint from P as well as from S_1 and S_2. Therefore, $(Z_i)_Q$ is a further blow-up along the strict transform of S_i.

Next suppose that ι is the standard quadratic embedding of $\mathbb{P}^1 \times \mathbb{P}^1$ into $\mathbb{P}^3$. In this case, $S_i \simeq \mathbb{P}^2$ are linear subspaces of $\mathbb{P}^4$ contained in Y, Z_Q is the blow-up at finitely many smooth points of a quadric cone and $(Z_i)_Q$ is the blow-up of Z_Q along one of the rays of the quadratic cone that miss the centers of the other blow-ups. Therefore $(Z_1)_Q \simeq (Z_2)_Q$, but this isomorphism does not extend to an isomorphism of Z_1 and Z_2.

This leads to a moduli space that is non-separated in a quite peculiar way: the point corresponding to the class of $(Z_1)_Q \simeq (Z_2)_Q$ completes the curve $\mathbb{P}^1 \setminus \{Q\}$ corresponding to the family $(Z_1)_{\mathbb{P}^1 \setminus \{Q\}} \simeq (Z_2)_{\mathbb{P}^1 \setminus \{Q\}}$ in two different way.

The result is the following 1-dimensional algebraic space. Let $Q \in \mathbb{P}^1$ a point. Take two copies of this $\mathbb{P}^1$ and glue them together along $\mathbb{P}^1 \setminus \{Q\}$. Then glue the two copies of Q together but by a separate gluing. Therefore there are two separate ways to get to Q from the rest of the $\mathbb{P}^1$.

As we mentioned before, a result of Matsusaka and Mumford tells us that in our case these pathologies do not occur.

Theorem 13.22 [MM64, Theorem 1]. *Let R be a DVR and $T = \operatorname{Spec} R$ with closed point $t_s \in T$. Further let X_1/T be a proper T-scheme and X_2/T a reduced T-scheme of finite type such that $(X_2)_{t_s}$ is not ruled. Assume that X_1 and X_2 are birational. Then so are $(X_1)_{t_s}$ and $(X_2)_{t_s}$.*

We may use this result to prove separatedness of $\mathcal{M}_h^{\text{smooth}}$, but first we need an auxiliary theorem.

Theorem 13.23. *Let S be a scheme and $f_i : X_i \to S$ two proper S-schemes, $\mathcal{L}_i$ relatively ample line bundles on X_i/S and $j_i : U_i \hookrightarrow X_i$ open immersions with complement $Z_i = X_i \setminus U_i$ for $i = 1, 2$. Assume that*

(13.23.1) there exists an S-isomorphism $\alpha : U_1/S \xrightarrow{\simeq} U_2/S$ such that $\alpha^* \mathcal{L}_2 \simeq \mathcal{L}_1$, and

(13.23.2) $\operatorname{depth}_{Z_i} X_i \geq 2$ for $i = 1, 2$ (this is satisfied if for example X_i is normal and $\operatorname{codim}(Z_i, X_i) \geq 2$).

Then α extends to X_1 to give an isomorphism $X_1/S \simeq X_2/S$.

Proof. Once α has an extension to X_1, it is unique, so the question is local on S and thus we may assume that it is affine. Let m be large enough that $\mathcal{L}_i^{\otimes m}$ is relatively very ample. First observe that (13.23.2) implies that $(j_i)_* \left(\mathcal{L}_i^{\otimes m}|_{U_i} \right) \simeq \mathcal{L}_i^{\otimes m}$ for $i = 1, 2$. Therefore

$$\left(f_i|_{U_i} \right)_* \left(\mathcal{L}_i^{\otimes m}|_{U_i} \right) \simeq (f_i)_* \mathcal{L}_i^{\otimes m}|_{f_i(U_i)}.$$

This implies that $\left(f_i|_{U_i}\right)_*\left(\mathscr{L}_i^{\otimes m}|_{U_i}\right)$ is coherent. Let $\mathscr{A}$ be an ample line bundle on S, then $\left(f_i|_{U_i}\right)_*\left(\mathscr{L}_i^{\otimes m}|_{U_i}\right)\otimes\mathscr{A}^{\otimes r}$ is generated by global sections for $r\gg 0$. As $\mathscr{L}_i^{\otimes m}$ is relatively very ample, this gives a surjection

$$f_i^*\left(\bigoplus^r \mathscr{A}^{-1}\right)\twoheadrightarrow \mathscr{L}_i^{\otimes m},$$

which in turn induces an embedding $\phi_i : X_i \hookrightarrow \mathbb{P}_S^{r-1}$.

As the isomorphism α in (13.23.1) gives an isomorphism between the sheaves

$$\left(f_1|_{U_1}\right)_*\left(\mathscr{L}_1^{\otimes m}|_{U_1}\right)\simeq\left(f_2|_{U_2}\right)_*\left(\mathscr{L}_2^{\otimes m}|_{U_2}\right),$$

we may choose the generators defining the ϕ_i to be compatible with this isomorphism and conclude that $\phi_1|_{U_1}=\phi_2|_{U_2}\circ\alpha$.

Since U_i is dense in X_i, we obtain that $\phi_i(X_i)$ is the Zariski closure of $\phi_i(U_i)$ and hence we have

$$X_1 \xrightarrow[\phi_1]{\simeq} \phi_1(X_1)=\overline{\phi_1(U_1)}=\overline{\phi_2(U_2)}=\phi_2(X_2)\xleftarrow[\phi_2]{\simeq} X_2.$$

Clearly, this isomorphism restricted to U_1 coincides with α proving the desired statement. □

Corollary 13.24. *$\mathcal{M}_h^{\text{smooth}}$ is separated.*

Proof. Let R be a DVR and $T=\operatorname{Spec} R$ with general point $t_g\hookrightarrow T$. Further let $(X_i\to T,\mathscr{L}_i)\in\mathcal{M}_h^{\text{smooth}}(T)$ two families for $i=1,2$ and assume that there exists an isomorphism $\alpha_g : ((X_1)_{t_g},(\mathscr{L}_1)_{t_g})\to((X_2)_{t_g},(\mathscr{L}_2)_{t_g})$.

Let $U_i\subseteq X_i$ be the largest open sets for $i=1,2$ such that there exists an extension of α_g that gives an isomorphism $\alpha : U_1\to U_2$.

Now observe that as α_g induces a birational equivalence between X_1 and X_2, by (13.22) it extends to a birational equivalence between $(X_1)_{t_s}$ and $(X_2)_{t_s}$ and hence these contain isomorphic open sets, which are then contained in U_1 and U_2 respectively. Therefore the conditions of (13.23) are satisfied and so α extends to an isomorphism $X_1/T\simeq X_2/T$. □

With this we have covered the most important properties of moduli functors, boundedness, local closedness, and separatedness. These properties, along with weak positivity and weak stability (see [Vie95, 7.16]) and projectivity of complete moduli (see [Kol90a] and (13.E)), allow one to prove the following:

Theorem 13.25 [Kol90a], [Vie95, 1.11]. *There exists a quasi-projective coarse moduli scheme for $\mathcal{M}_h^{\text{smooth}}$.*

For more precise statements see [Kol90a] and [Vie95, §1.2]. Other relevant sources are [Kol85, KSB88, Vie89, Vie90a, Vie90b, Vie06].

At first sight it may seem that with the construction of this moduli scheme we have accomplished the plan laid down earlier. However, it is not entirely so. We should definitely consider this an answer if we only care about smooth canonically polarized varieties. After all, the moduli space does "classify" these objects. On the other hand, a canonical model may not be smooth. So if we also care about those cases, we have to work with singular varieties as well. Moreover, in order to obtain a compact moduli space we have to allow singular varieties:

DEFINITION 13.26. A variety X is called **stable** if X is semi-log canonical, projective and ω_X is an ample $\mathbb{Q}$-line bundle.

Extending the definition of $\mathcal{M}_h^{\text{smooth}}$ to stable varieties is not entirely straightforward because working with $\mathbb{Q}$-line bundles and not necessarily normal singularities requires some extra care. We will discuss some of the relevant issues in Chapter 14.

Once the moduli functor is defined, we would like to have the same properties for this extended functor as we did for $\mathcal{M}_h^{\text{smooth}}$. Unfortunately, the proof of (13.24) does not extend to this case, so we need a new approach to prove that the extended moduli functor is also separated.

Kollár and Shepherd-Barron showed that the minimal model program provides a way to do this [KSB88]. The main idea is the following: Consider a family $\phi : X \to B$ of varieties of general type and let $\psi : Y \to B$ be its ltm over B cf. (5.57). Then by the Basepoint-free theorem (5.1) we may consider the canonical model $\zeta : Z \to B$ over B. If the general fiber of ϕ is a smooth canonically polarized variety, then the same is true for ψ and ζ. Therefore this should be considered the typical family of stable varieties whose general fiber is smooth. It follows from the mmp that the special fibers are also stable. In fact this method works in full generality and was used by Kollár and Shepherd-Barron to prove separatedness of the moduli functor of stable surfaces. For more details on this line of argument and how separatedness of the moduli functor of stable surfaces follows from this, see [KSB88, §5].

13.E Properness

A moduli functor being proper is not necessary for constructing a moduli space that (coarsely) represents it, but it is necessary for that moduli space to be proper.

DEFINITION 13.27. Let $\mathcal{F}$ be a subfunctor of $\mathcal{MP}$. $\mathcal{F}$ is said to be **proper** or **complete** if it is separated, bounded and the following condition is satisfied:

Let T be the spectrum of a DVR with generic point $\iota : t_g \to T$ and

$$(f_g : X_g \to T_g, \mathcal{L}_g) \in \mathcal{F}(T_g).$$

Then there exist a spectrum of a DVR, $\widetilde{T}$ with generic point $\widetilde{\iota} : \widetilde{t}_g \to \widetilde{T}$, a finite and dominant morphism $\tau : \widetilde{T} \to T$ and

$$(\widetilde{f} : \widetilde{X} \to \widetilde{T}, \widetilde{\mathcal{L}}) \in \mathcal{F}(\widetilde{T})$$

such that

$$\tau^*(f_g : X_g \to T_g, \mathscr{L}_g) \simeq \iota^*(\widetilde{f} : \widetilde{X} \to \widetilde{T}, \widetilde{\mathscr{L}}).$$

In other words every family $(f_g : X_g \to T_g, \mathscr{L}_g)$ can be extended to a family over the closed point after a finite base change.

As mentioned above $\mathcal{M}_h^{\text{smooth}}$ is **not** proper. This is exactly why we need to extend this functor to include stable varieties. An important ingredient in proving properness of the extended funtor is the mmp. For more on this subject see [KSB88], [Kol90b], [Kol].

CHAPTER 14

Families and moduli functors

A very important issue in considering higher dimensional moduli problems is that, as opposed to the case of curves, when studying families of higher dimensional varieties one must put conditions on the admissible families that restrict the kind of families and not only the kind of fibers that are allowed. This is perhaps better understood through an example of bad behavior.

14.A An important example

- Let $R \subseteq \mathbb{P}^4$ be a quartic rational normal curve, i.e., the image of the embedding of $\mathbb{P}^1$ into $\mathbb{P}^4$ by the global sections of $\mathscr{O}_{\mathbb{P}^1}(4)$. For example take

$$R = \{[u^4 : u^3v : u^2v^2 : uv^3 : v^4] \in \mathbb{P}^4 \mid [u : v] \in \mathbb{P}^1\}.$$

- Let $T \subseteq \mathbb{P}^5$ be a quartic rational scroll, i.e., the image of the embedding of $\mathbb{P}^1 \times \mathbb{P}^1$ into $\mathbb{P}^5$ by the global sections of $\mathscr{O}_{\mathbb{P}^1\times\mathbb{P}^1}(1,2)$. Let f_1 and f_2 denote the divisor classes of the two rulings on T. For example take

$$T = \{[xz^2 : xzt : xt^2 : yz^2 : yzt : yt^2] \in \mathbb{P}^5 \mid ([x : y], [z : t]) \in \mathbb{P}^1 \times \mathbb{P}^1\}.$$

- Let $C_R \subseteq \mathbb{P}^5$ be the projectivized cone over R in $\mathbb{P}^5$ and $C_T \subseteq \mathbb{P}^6$ the projectivized cone over T in $\mathbb{P}^6$. For the above choices, these are represented by

$$C_R = \{[u^4 : u^3v : u^2v^2 : uv^3 : v^4 : w^4] \in \mathbb{P}^5 \mid [u : v : w] \in \mathbb{P}^2\}, \text{ and}$$
$$C_T = \{[xz^2 : xzt : xt^2 : yz^2 : yzt : yt^2 : pq^2] \in \mathbb{P}^6 \mid$$
$$\mid ([x : y : p], [z : t : q]) \in \mathbb{P}^2 \times \mathbb{P}^2 \setminus (\ell_1 \cup \ell_2)\},$$

where $\ell_1 = (p = z = t = 0), \ell_2 = (x = y = q = 0) \subseteq \mathbb{P}^2 \times \mathbb{P}^2$ are two lines. Notice also that the parametrization of C_T given above is not generically finite.

- Let $V \subseteq \mathbb{P}^5$ be a Veronese surface, i.e., the image of the Veronese embedding; the embedding of $\mathbb{P}^2$ into $\mathbb{P}^5$ by the global sections of $\mathscr{O}_{\mathbb{P}^2}(2)$. For example take

$$V = \{[u^2 : vw : uv : uw : v^2 : w^2] \mid [u : v : w] \in \mathbb{P}^2\}.$$

Another possible parameterization is obtained when the Veronese embedding is combined with the 4-to-1 endomorphism of $\mathbb{P}^2$, $[u : v : w] \mapsto [u^2 : v^2 : w^2]$:

$$V = \{[u^4 : v^2w^2 : u^2v^2 : u^2w^2 : v^4 : w^4] \mid [u : v : w] \in \mathbb{P}^2\}.$$

- Let $W \subseteq \mathbb{P}^5 \times \mathbb{A}^1$ be the following quasi-projective threefold:

$$W = \Big\{\big([u^4 : u^3v + \xi(v^2w^2 - u^3v) : u^2v^2 : uv^3 + \xi(u^2w^2 - uv^3) : v^4 : w^4], \xi\big)\Big| \\ \Big|[u : v : w] \in \mathbb{P}^2, \xi \in \mathbb{A}^1\Big\} \subseteq \mathbb{P}^5 \times \mathbb{A}^1.$$

OBSERVATIONS:

- *V is a smoothing of C_R.* Indeed, the second projection of $\mathbb{P}^5 \times \mathbb{A}^1$ exhibits W as a family of surfaces $W \to \mathbb{P}^1$. Both C_R and V appear as members of this family. For $\xi = 0, 1 \in \mathbb{A}^1$; $W_0 \simeq C_R$ and $W_1 \simeq V$.

- *R is a hyperplane section of T.* Indeed let $H \subseteq \mathbb{P}^5$ be a general hyperplane. Then $C := H \cap T$ is a smooth curve such that $C \sim \mathfrak{f}_1 + 2\mathfrak{f}_2$. Then by the adjunction formula $2g(C) - 2 = (-2\mathfrak{f}_1 - 2\mathfrak{f}_2 + C) \cdot C = -2$, hence $C \simeq \mathbb{P}^1$. Furthermore, then $C^2 = 4$, so $\mathscr{O}_T(1,2)\big|_C \simeq \mathscr{O}_C(4)$. Therefore C is a quartic rational curve in $H \simeq \mathbb{P}^4$, and thus it may be identified with R.

- *T is also a smoothing of C_R.* Indeed, both T and C_R are hyperplane sections of C_T. The latter statement follows from the previous observation.

ANALYSIS:

- It is relatively easy, and thus left to the reader, to compute that C_R has log terminal singularities. In particular, this type of singularity is among those we have to be able to handle.

- The problem this example points to is that if we allow arbitrary families, then we may get unwanted results. For example, using the families derived from C_T and W would mean that $T \simeq \mathbb{P}^1 \times \mathbb{P}^1$ and $V \simeq \mathbb{P}^2$ should be considered to have the same deformation type. However, there are obviously no smooth families that they both belong to, as they are topologically very different. For instance, $K_T^2 = 8$ while $K_V^2 = 9$.

- The crux of the matter is that C_T is *not* $\mathbb{Q}$-Gorenstein (cf. (3.5)) and consequently the family obtained from it is not a $\mathbb{Q}$-Gorenstein family. This is actually an important point: the members of the family are $\mathbb{Q}$-Gorenstein surfaces, but the relative canonical bundle of the family is not $\mathbb{Q}$-Cartier. In particular, the canonical divisors of the members of the family are not consistent.
- The family obtained from W *is* $\mathbb{Q}$-Gorenstein and consequently the canonical divisors of the members of the family are similar to some extent. Among other things this implies that $K_{C_R}^2 = 9$. One may also use the parameterization of C_R given above to verify this fact independently. It is interesting to note that K_{C_R} is $\mathbb{Q}$-Cartier, but not Cartier even though its self-intersection number is an integer.

14.B $\mathbb{Q}$-Gorenstein families

We have seen that we have to extend the definition of the moduli functor (see (11.5)) to allow (some) singular varieties.

The reader should be warned that here we are entering a somewhat uncharted territory. This area of research is still rapidly evolving. In particular it has not yet crystallized what the "right" or optimal conditions to assume in the definition of the moduli functor are. There are several different moduli functors that are being studied at this time. Accordingly, on occasion, we may assume too much or too little. This chapter is intended to give a peek into the forefront of the research that is conducted in this area.

The example in (14.A) shows that it is not enough to restrict the kind of members of the families we allow but we have to restrict the kind of families we allow as well.

DEFINITION 14.1. Let k be an algebraically closed field of characteristic 0 and Sch_k the category of k-schemes. We define $\mathcal{M}^{\mathrm{wst}} : \mathsf{Sch}_k \to \mathsf{Sets}$, the **moduli functor of weakly stable canonically polarized $\mathbb{Q}$-Gorenstein varieties**, the following way.

(14.1.1) A morphism $f : X \to B$ of schemes is called a **weakly stable family** if the following hold:

(a) f is flat and projective with connected fibers,

(b) $\omega_{X/B}$ is a relatively ample $\mathbb{Q}$-line bundle, and

(c) for all $b \in B$, X_b has only semi-log canonical singularities.

(14.1.2) For an object $B \in \operatorname{Ob} \mathsf{Sch}_k$,

$$\mathcal{M}^{\mathrm{wst}}(B) := \{f : X \to B \mid f \text{ is a weakly stable family}\} \big/ \simeq$$

where "$\simeq$" is defined as in (11.5).

(14.1.3) For any morphism $\alpha \in \operatorname{Hom}_{\mathsf{Sch}_k}(A, B)$,

$$\mathcal{M}^{\mathrm{wst}}(\alpha) := (\_\_) \times_B \alpha.$$

REMARK 14.2. Note that it is not obvious from the definition that this is indeed a functor. However, this functor (if it is a functor) is actually not yet the one we are interested in.

We will use this to define others. The fact that those others are indeed functors will follow from Lemma 14.4.

As mentioned above, this functor is not yet the right one. There are two additional conditions to which we have to pay attention. The first is to keep track of the Hilbert polynomials of the polarizations. This is straightforward, although somewhat different from the smooth case in that now we have to also keep track of what power of the $\mathbb{Q}$-line bundle corresponds to the polarization. This is done as follows.

DEFINITION 14.3. Let k be an algebraically closed field of characteristic 0, Sch_k the category of k-schemes and $N \in \mathbb{N}$. We define $\mathcal{M}^{\mathrm{wst},[N]} : \mathsf{Sch}_k \to \mathsf{Sets}$, the **moduli functor of weakly stable canonically polarized $\mathbb{Q}$-Gorenstein varieties of index N**, as the subfunctor of $\mathcal{M}^{\mathrm{wst}}$ with the additional condition that $\omega_{X/B}^{[N]}$ is a line bundle:

$$\mathcal{M}^{\mathrm{wst},[N]}(B) := \left\{ (f : X \to B) \in \mathcal{M}^{\mathrm{wst}}(B) \mid \omega_{X/B}^{[N]} \text{ is a line bundle} \right\}.$$

Now let $h \in \mathbb{Q}[t]$. Then

$$\mathcal{M}_h^{\mathrm{wst},[N]}(B) := \left\{ (f : X \to B) \in \mathcal{M}^{\mathrm{wst},[N]}(B) \mid \chi(X_b, \omega_{X_b}^{[mN]}) = h(m) \right\}.$$

In order to use the polarization given by the appropriate reflexive power of the canonical sheaves of the fibers, we need to know that the powers of the relative canonical sheaf commute with base change. The following shows that for objects in $\mathcal{M}^{\mathrm{wst},[N]}(B)$, this holds for multiples of the index.

Lemma 14.4 [HK04, 2.6]. *Given a weakly stable family of canonically polarized $\mathbb{Q}$-Gorenstein varieties of index N, $f : X \to B$, and a morphism $\alpha : T \to B$, we have*

$$\alpha_X^* \omega_{X/B}^{[N]} \simeq \omega_{X_T/T}^{[N]}.$$

Proof. Let $U \subset X$ be the largest open subset U of X such that ω_{U_b} is a line bundle for all $b \in B$ or equivalently the largest open subset U of X such that $\omega_{X/B}|_U \simeq \omega_{U/B}$ is a line bundle. Then $\omega_{X/B}^{[N]}|_U \simeq \omega_{U/B}^{\otimes N}$ and hence

$$\alpha_X^* \omega_{X/B}^{[N]}|_{\alpha_X^{-1}U} \simeq \alpha_X^* \omega_{U/B}^{\otimes N} \simeq \omega_{\alpha_X^{-1}U/T}^{\otimes N} \simeq \omega_{X_T/T}^{[N]}|_{\alpha_X^{-1}U}.$$

Now $\operatorname{codim}(U_b, X_b) \geq 2$ for all $b \in B$ (cf. (3.14)), so $\operatorname{codim}((\alpha_X^{-1}U)_t, (X_T)_t) \geq 2$ for all $t \in T$ and hence $\operatorname{codim}(\alpha_X^{-1}U, X_T) \geq 2$. Finally $\alpha_X^* \omega_{X/B}^{[N]}$ and $\omega_{X_T/T}^{[N]}$ are reflexive, so since they are isomorphic on $\alpha_X^{-1}U$, they are isomorphic on X_T. □

However, there is an issue that complicates the matter. It is possible that the information carried by the functor $\mathcal{M}^{\mathrm{wst},[N]}$ is not enough to encode the main topological properties of the fibers. As a solution, Kollár suggests that we require more.

DEFINITION 14.5 : KOLLÁR'S CONDITION. We say that **Kollár's condition** holds for any family $(f : X \to B) \in \mathcal{M}^{\mathrm{wst}}(B)$, if for all $\ell \in \mathbb{Z}$ and for all $b \in B$,

$$\omega_{X/B}^{[\ell]}|_{X_b} \simeq \omega_{X_b}^{[\ell]}.$$

The important difference between this condition and the situation in the previous lemma is that this condition requires that the restriction of *all* reflexive powers commute with base change, not only those that are line bundles.

It is relatively easy to see using the same argument as in the proof of (14.4) that this condition is equivalent to the requirement that the restriction of all reflexive powers to the fibers be reflexive themselves.

Now we are ready to define the "right" moduli functor.

DEFINITION 14.6. Let k be an algebraically closed field of characteristic 0 and Sch_k the category of k-schemes. We define $\mathcal{M} = \mathcal{M}^{\mathrm{st}} : \mathsf{Sch}_k \to \mathsf{Sets}$, the **moduli functor of stable canonically polarized $\mathbb{Q}$-Gorenstein varieties**, as the subfunctor of $\mathcal{M}^{\mathrm{wst}}$ with the additional condition that a family $(f : X \to B) \in \mathcal{M}^{\mathrm{wst}}(B)$ satisfy Kollár's condition:

$$\mathcal{M}(B) := \left\{ (f : X \to B) \in \mathcal{M}^{\mathrm{wst}}(B) \mid \forall \ell \in \mathbb{Z}, b \in B, \omega_{X/B}^{[\ell]}\big|_{X_b} \simeq \omega_{X_b}^{[\ell]} \right\}.$$

Finally, let $h \in \mathbb{Q}[t]$ and $N \in \mathbb{N}$. Then we define $\mathcal{M}_h^{[N]}$ as the subfunctor of $\mathcal{M}^{[N]}$ with the additional condition that for any family $(f : X \to B) \in \mathcal{M}^{[N]}(B)$, the Hilbert polynomial of the fibers agree with h:

$$\mathcal{M}_h^{[N]}(B) := \left\{ (f : X \to B) \in \mathcal{M}^{[N]}(B) \mid \forall b \in B, \chi(X_b, \omega_{X_b}^{[mN]}) = h(m) \right\}.$$

The difference between the moduli functors $\mathcal{M}_h^{\mathrm{wst},[N]}$ and $\mathcal{M}_h^{[N]}$ is very subtle. They parameterize the same objects and as long as one restricts to Gorenstein varieties, they allow the same families. This means that if one is only interested in the compactification of the coarse moduli space of $\mathcal{M}_h^{\mathrm{smooth}}$, then the difference between these two moduli functors does not matter as they lead to the same reduced scheme. The difference may only show up in their scheme structure. However, the usefulness of a moduli space is closely related to its "right" scheme structure, so it is important to find the correct one.

A somewhat troubling point is that we do not actually know for a fact that these two moduli functors are really different in characteristic 0. In other words, we do not know an example of a family that belongs to $\mathcal{M}_h^{\mathrm{wst},[N]}$, but not to $\mathcal{M}_h^{[N]}$. The following example of Kollár shows that these functors *are* different in characteristic $p > 0$, but there is no similar example known in characteristic 0.

14.7 KOLLÁR'S EXAMPLE. Note that the first part of the discussion (14.7.1) works in arbitrary characteristic. It shows that a family with the required properties belongs to $\mathcal{M}_h^{\mathrm{wst},[N]}$, but not to $\mathcal{M}_h^{[N]}$. In the second part (14.7.2) it is shown that in characteristic $p > 0$ a family satisfying another set of properties also has the ones required in (14.7.1). Finally, it is easy to see that the example in (14.A) admits these later properties, so we do indeed have an explicit example for this behavior.

(14.7.1) Suppose that $g : Y \to B$ is a family of canonically polarized $\mathbb{Q}$-Gorenstein varieties (with only semi-log canonical singularities) and assume that $B = \operatorname{Spec} R$ with $R = (R, \mathfrak{m})$ a DVR. Let $B_n = \operatorname{Spec} R_n$ where $R_n := R/\mathfrak{m}^n$ and consider the restriction of the family g over B_n, $g_n : Y_n = Y \times_B B_n \to B_n$. Finally assume that ω_{Y_n/B_n} is

$\mathbb{Q}$-Cartier of index r_n for all n but $r_n \to \infty$ as $n \to \infty$ (recall that the index means the smallest integer m such that the m-th reflexive power is a line bundle). Note that by Lemma 14.4 this implies that $\omega_{Y/B}$ cannot be $\mathbb{Q}$-Cartier.

We claim that g_n is a weakly stable family of canonically polarized $\mathbb{Q}$-Gorenstein varieties of index r_n (Definition 14.1), but it does not satisfy Kollár's condition (Definition 14.5) for all but possibly a finite number of n.

The first part of the claim is obvious from the assumptions. For the second part consider the following argument. If g_n satisfied Kollár's condition, then for any $m < n$ the restriction of $\omega_{Y_n/B_n}^{[r_m]}$ to Y_m (hence to Y_1) would be a line bundle implying, via Nakayama's lemma, that $\omega_{Y_n/B_n}^{[r_m]}$ itself is a line bundle. That however would further imply that $r_n \leq r_m$, but since $r_n \to \infty$ as $n \to \infty$, this can only happen for a finite number of n's.

(14.7.2) Next we will show (following Kollár) that a family such as in (14.7.1) does exist in characteristic $p > 0$. It is currently not known whether such an example exists in characteristic 0. As above, let $g : Y \to B$ be a family of canonically polarized $\mathbb{Q}$-Gorenstein varieties with only log canonical singularities, such that $B = \operatorname{Spec} R$ with $R = (R, \mathfrak{m})$ a DVR. Assume that g, Y, and B are defined over a field k of characteristic $p > 0$. Let $B_n = \operatorname{Spec} R_n$ where $R_n := R/\mathfrak{m}^n$ and consider the restriction of the family g over B_n, $g_n : Y_n = Y \times_B B_n \to B_n$. For a concrete example one may consider the smoothing of C_R to T via C_T (reduced over $k[x]_{(x)}$) from the example in (14.A).

Claim. ω_{Y_n/B_n} is $\mathbb{Q}$-Cartier.

Proof. The question is local on Y, so we may assume that Y_n is a local scheme. In particular, we will assume that all line bundles on Y_1 are trivial. Let

$$\iota_n : U_n = (Y_n \setminus \operatorname{Sing} Y_n) \hookrightarrow Y_n.$$

By assumption, Y_n is normal (R_1 and S_2) for all n, so

$$\omega_{Y_n/B_n}^{[m]} \simeq (\iota_n)_* \omega_{U_n/B_n}^{\otimes m}$$

for all m.

Next, consider the restriction maps to the special fiber of the family from all the infinitesimal thickenings:

$$\varrho_n : \operatorname{Pic} U_n \to \operatorname{Pic} U_1.$$

The key observation is the following: the kernel of this map is a (p-power) torsion group (cf. [Har77, Ex.III.4.6]). In other words, any line bundle on U_n whose restriction to U_1 is trivial extends to a $\mathbb{Q}$-Cartier divisor on Y_n.

Recall that by assumption, $\omega_{Y_1/B_1} = \omega_{Y_1}$ is $\mathbb{Q}$-Cartier (of index r_1), in particular $\omega_{Y_1/B_1}^{[r_1]}$ is trivial. Therefore,

$$\varrho_n(\omega_{U_n/B_n}^{\otimes r_1}) = \omega_{U_n/B_n}^{\otimes r_1}\big|_{U_1} \simeq \omega_{U_1/B_1}^{\otimes r_1}$$

is also trivial. Consequently,

$$\omega_{U_n/B_n}^{\otimes r_1} \in \ker \varrho_n.$$

Recall that this is a torsion group, so there exists an $m_n \in \mathbb{N}$ such that $\left(\omega_{U_n/B_n}^{\otimes r_1}\right)^{\otimes m_n}$ is trivial. That however, implies that then so is

$$\omega_{Y_n/B_n}^{[r_1 \cdot m_n]} \simeq (\iota_n)_* \omega_{U_n/B_n}^{\otimes (r_1 \cdot m_n)} \simeq (\iota_n)_* \mathcal{O}_{U_n} \simeq \mathcal{O}_{Y_n}.$$

We conclude that ω_{Y_n/B_n} is indeed $\mathbb{Q}$-Cartier. □

It is left for the reader to prove that if $\omega_{Y/B}$ is not $\mathbb{Q}$-Cartier, then the index of ω_{Y_n/B_n} has to tend to infinity. It is easy to check that this happens in the case of C_T considered as a non-$\mathbb{Q}$-Gorenstein smoothing of C_R as above.

REMARK 14.8. The previous example also shows an important aspect of why Kollár's condition is useful. Let Z be a canonically polarized $\mathbb{Q}$-Gorenstein variety of index m with only semi-log canonical singularities. If we want to find a moduli space where Z appears, we may choose the moduli functor $\mathcal{M}_h^{\text{wst},[a\cdot m]}$ for any $a \in \mathbb{N}$ (where h is the Hilbert polynomial of $\omega_Z^{[a\cdot m]}$). The previous example shows that the scheme structure of the corresponding moduli space will depend on which a we choose. As a grows, the moduli scheme gets thicker. Consequently, there is not a unique moduli scheme where Z would naturally belong. This does not happen for the functor $\mathcal{M}_h^{[a\cdot m]}$ because Kollár's condition makes sure that the choice of a makes no difference.

14.C Projective moduli schemes

With the definition of $\mathcal{M}_h^{[N]}$ we have reached the moduli functor that should be the right one. This functor accounts for all canonical models, even a little bit more, as well as all degenerations of smooth canonical models.

The natural next step would be to state the equivalent of Theorem 13.25 for $\mathcal{M}_h^{[N]}$. However, we can't quite do that exactly.

Boundedness was proven for moduli of surfaces (i.e., $\deg h = 2$) in [Ale94] (cf. [AM04]) A more general result was obtained in [Kar00] assuming that certain conjectures from the Minimal Model Program were true. Fortunately these conjectures have been recently proven in [HM07, BCHM10] and explained in Part II of this book, so this piece of the puzzle is in place.

Separatedness follows from [KSB88] and [Kaw07].

Projectivity follows from [Kol90a].

Local closedness for $\mathcal{M}_h^{\text{wst},[N]}$ was proven in [HK04]. Hacking obtained partial results toward the local closedness of $\mathcal{M}_h^{[N]}$ in [Hac04]. Local closedness of $\mathcal{M}_h^{[N]}$ in general has been proven by Abramovich and Hassett, but this result has not appeared in any form yet at the time of this writing. Even more recently a general flattening result that implies the local closedness of $\mathcal{M}_h^{[N]}$ has been proven by Kollár [Kol08a]. Kollár's result essentially closes the question of local closedness for good.

So, the conclusion is that all the pieces are in place, even though the statement of the existence of a projective coarse moduli scheme for $\mathcal{M}_h^{[N]}$ has not yet appeared in print and thus we will not formulate it as a theorem here.

14.D Moduli of pairs and other generalizations

As it has become clear in higher dimensional geometry in recent years, the "right" formulation of (higher dimensional) problems deals with pairs, or log varieties (cf. [Kol97b]). Accordingly, one would like to have a moduli theory of log varieties. In fact, one would like to go through the entire Part III of this book and replace all objects with log varieties, canonical models with log canonical models, etc.

However, this is not as straightforward as it may appear at first sight and the formulation of the moduli functor itself is not entirely obvious. The good news is that as it is shown in Part II, the necessary tools from the minimal model program are now available. Steady work is being done in this area and perhaps by the time these words appear in print, there will be concrete results to speak of about log varieties. For partial results in this direction the reader should consult recent works of János Kollár [Kol07c, Kol08a, Kol08b, Kol08c, Kol09]. A different, but very promising approach has been taken by Dan Abramovich and Brendan Hassett in [AH09].

There are many related results we did not have the chance to mention in detail. Here is a somewhat random sample of those results: Valery Alexeev has been particularly prolific and the interested reader should take a look at his results, a good chunk of which is joint work with Michel Brion: [Ale96, Ale02, Ale01, AB04a, AB04b, AB05]. A few years ago Paul Hacking solved the long-standing problem of compactifying the moduli space of plane curves in a geometrically meaningful way [Hac04]. Recently Hacking, jointly with Sean Keel and Jenia Tevelev, has done the same for the moduli space of hyperplane arrangements [HKT06] and Del Pezzo surfaces [HKT07].

In the last two chapters we will take a look at two special topics. First we will review some recent results on the singularities of stable varieties and then some results on the geometry of their moduli spaces.

CHAPTER 15

Singularities of stable varieties

In this chapter we will recall a few results regarding singularities that occur on stable varieties, that is, singularities of the objects that appear on the boundary of the moduli spaces we discussed in the previous chapter.

We have already defined rational and DB singularities cf. (3.52) and (3.57) and here we will review some of their basic properties and some recent results connecting them to the singularities of the mmp. As an application of the main results we will investigate the following question: If one is only interested in stable degenerations of smooth canonically polarized varieties, in other words in the connected components of the moduli space that contain moduli points representing smooth varieties, then is there any additional assumption one might make about the singularities that appear? It is easy to see that there are certain properties one cannot expect to keep, such as for example being *normal* or *Gorenstein*. However, we will see that being Cohen-Macaulay is preserved in stable families.

We will need the following definition and foundational results: Let X be a complex scheme of finite type and let $\omega_X^{\bullet}$ denote the **dualizing complex** of X, i.e., $\omega_X^{\bullet} = f^!\mathbb{C}$, where $f : X \to \operatorname{Spec}\mathbb{C}$ is the structure map cf. [Har66]. As introduced in (3.54), the symbol $\simeq_{\text{qis}}$ stands for quasi-isomorphism of complexes, which is isomorphism in the derived category setting. The next two theorems essential statement is that $\mathcal{R}\phi_*$ has both a right $\phi^!$ and a left $\mathcal{L}\phi^*$ adjoint.

Theorem 15.1 [Har66, VII] Grothendieck duality. *Let $\phi : Y \to X$ be a proper morphism, then for all $G^{\bullet}$ bounded complexes of $\mathcal{O}_Y$-modules with coherent cohomology sheaves,*

$$\mathcal{R}\phi_*\mathcal{R}\mathcal{H}om_Y(G^{\bullet}, \omega_Y^{\bullet}) \simeq_{\text{qis}} \mathcal{R}\mathcal{H}om_X(\mathcal{R}\phi_*G^{\bullet}, \omega_X^{\bullet}).$$

Theorem 15.2 [Har66, II.5.10] Adjointness. *Let $\phi : Y \to X$ be a proper morphism, then for all $F^{\bullet}$ bounded complexes of $\mathcal{O}_X$-modules and $G^{\bullet}$ bounded complexes of $\mathcal{O}_Y$-*

modules with coherent cohomology sheaves,

$$\mathcal{R}\phi_* \mathcal{RH}om_Y(\mathcal{L}\phi^* F^\bullet, G^\bullet) \simeq_{\text{qis}} \mathcal{RH}om_X(F^\bullet, \mathcal{R}\phi_* G^\bullet).$$

15.A Singularity criteria

The moral of the theorems in this section can be summarized by the following principle:

The splitting principle 15.3. *Morphisms do not split accidentally.*

REMARK 15.4. It is customary to casually use the word "splitting" to explain the statements of the theorems that follow. However, the reader should be warned that one has to be careful with the meaning of this, because these "splittings" take place in the derived category, which is not abelian. For this reason, in the statements of the theorems below we use the terminology that a morphism admits a *left inverse*. In an abelian category this condition is equivalent to "splitting" and being a direct component (of a direct sum). With a slight abuse of language we labeled these as "Splitting theorems" cf. (15.5), (15.11) and (15.18).

The first theorem we recall is a criterion for a singularity to be rational.

Theorem 15.5 [Kov00b] Splitting theorem I. *Let $\phi : Y \to X$ be a proper morphism of varieties over $\mathbb{C}$ and $\varrho : \mathcal{O}_X \to \mathcal{R}\phi_*\mathcal{O}_Y$ the associated natural morphism. Assume that Y has rational singularities and ϱ has a left inverse, i.e., there exists a morphism (in the derived category of $\mathcal{O}_X$-modules) $\varrho' : \mathcal{R}\phi_*\mathcal{O}_Y \to \mathcal{O}_X$ such that $\varrho' \circ \varrho$ is a quasi-isomorphism of $\mathcal{O}_X$ with itself. Then X has only rational singularities.*

REMARK 15.6. Note that ϕ in the theorem does not have to be birational or even generically finite. It follows from the conditions that it is surjective.

Corollary 15.7. *Let X be a complex variety and $\phi : Y \to X$ a resolution of singularities. If $\mathcal{O}_X \to \mathcal{R}\phi_*\mathcal{O}_Y$ has a left inverse, then X has rational singularities.*

Corollary 15.8. *Let X be a complex variety and $\phi : Y \to X$ a finite morphism. If Y has rational singularities, then so does X.*

Using this criterion it is quite easy to prove that dlt singularities are rational. Here we only deal with the case when X is log terminal, i.e., (X, Δ) is klt and $\Delta = 0$. For the more general case see [KM98, 5.22] and the references therein.

Corollary 15.9 [Elk81]. *Let (X, Δ) be a dlt pair. Then X has rational singularities.*

Proof. By [KM98, 2.43] we may assume that (X, Δ) is a klt pair. Let $\phi : Y \to X$ be a log resolution. Since (X, Δ) has klt singularities,

$$\mathcal{O}_Y \hookrightarrow \mathcal{O}_Y(\lceil K_Y - \phi^*(K_X + \Delta)\rceil).$$

Consider the composition

$$\mathcal{O}_X \xrightarrow{\ \varrho\ } \mathcal{R}\phi_*\mathcal{O}_Y \xrightarrow{\ \varrho'\ } \mathcal{R}\phi_*\mathcal{O}_Y(\lceil K_Y - \phi^*(K_X + \Delta)\rceil) \simeq \mathcal{O}_X.$$

The isomorphism $\mathcal{R}\phi_*\mathcal{O}_Y(\lceil K_Y - \phi^*(K_X + \Delta)\rceil) \simeq \mathcal{O}_X$ follows from Kawamata-Viehweg vanishing (3.45) and the fact that $\lceil K_Y - \phi^*(K_X + \Delta)\rceil$ is an effective ϕ-exceptional divisor cf. [KMM87, 1-3-2]. The composition $\varrho' \circ \varrho$ is clearly an isomorphism and hence the statement follows from (15.5). □

REMARK 15.10. The fact that (15.5) implies (15.9) was first shown in [Kov00b]. The above proof is due to Karl Schwede cf. [ST07].

We also have a criterion for DB singularities that is similar to the one in (15.5):

Theorem 15.11 [Kov99, 2.3] Splitting theorem II. *Let X be a complex variety. If $\mathcal{O}_X \to \underline{\Omega}^0_X$ has a left inverse, then X has DB singularities.*

This criterion has several important consequences. We recall one of them now:

Corollary 15.12 [Kov99, 2.6]. *Let X be a complex variety with rational singularities. Then X has DB singularities.*

Proof. Let $\phi : Y \to X$ be a resolution of singularities. Then since Y is smooth the natural map $\varrho : \mathcal{O}_X \to \mathcal{R}\phi_*\mathcal{O}_Y$ factors through $\underline{\Omega}^0_X$ by (3.55.6). Then, since X has rational singularities, ϱ is a quasi-isomorphism, so we obtain that the natural map $\mathcal{O}_X \to \underline{\Omega}^0_X$ has a left inverse. Therefore, X has DB singularities by (15.11). □

Recently a few more critera have been found for DB singularities:

Theorem 15.13 [KSS10, 3.1]. *Let X be a normal Cohen-Macaulay scheme of finite type over $\mathbb{C}$. Let $\pi : X' \to X$ be a log resolution, and denote the reduced exceptional divisor of π by G. Then X has DB singularities if and only if $\pi_*\omega_{X'}(G) \simeq \omega_X$.*

Related results have been obtained in the non-normal Cohen-Macaulay case, see [KSS10] for details.

REMARK 15.14. The submodule $\pi_*\omega_{X'}(G) \subseteq \omega_X$ is independent of the choice of log resolution. Thus this submodule may be viewed as an invariant that partially measures how far a scheme is from being DB (compare with [Fuj08b]).

As an easy corollary, we obtain another proof that rational singularities are DB (this time via the Kempf-criterion for rational singularities).

Corollary 15.15 (15.12)**.** *Let X be a complex variety with rational singularities. Then X has DB singularities.*

Proof. Since X has rational singularities, it is Cohen-Macaulay and normal. Then $\pi_*\omega_{X'} = \omega_X$ but we also have $\pi_*\omega_{X'} \subseteq \pi_*\omega_{X'}(G) \subseteq \omega_X$, and thus $\pi_*\omega_{X'}(G) = \omega_X$ as well. The statement now follows from Theorem 15.13. □

We also see immediately that log canonical singularities coincide with DB singularities in the Gorenstein case.

Corollary 15.16. *Suppose that X is Gorenstein and normal. Then X is DB if and only if X is log canonical.*

Proof. X is easily seen to be log canonical if and only if $\pi_*\omega_{X'/X}(G) \simeq \mathcal{O}_X$. The projection formula then completes the proof. □

In fact, a slightly jazzed up version of this argument can be used to show that every Cohen-Macaulay log canonical pair is DB:

Corollary 15.17 [KSS10, 3.16]. *CM log canonical singularities are DB.*

We will see below that it is actually not necessary to assume CM in the previous theorem. However, the characterization of DB singularities in (15.13) is still useful on its own.

Theorem 15.18 [KK10] Splitting theorem III. *Let $f : Y \to X$ be a proper morphism between reduced schemes of finite type over $\mathbb{C}$. Let $W \subseteq X$ and $F := f^{-1}(W) \subset Y$ be closed reduced subschemes with ideal sheaves $\mathscr{I}_{W\subseteq X}$ and $\mathscr{I}_{F\subseteq Y}$. Assume that the natural map ϱ*

$$\mathscr{I}_{W\subseteq X} \xrightarrow[\varrho]{} \mathcal{R}f_*\mathscr{I}_{F\subseteq Y} \quad (\text{with left inverse } \varrho' \text{ drawn as a dashed arrow back})$$

admits a left inverse ϱ', that is, $\varrho' \circ \varrho = \mathrm{id}_{\mathscr{I}_{W\subseteq X}}$. Then if Y, F, and W all have DB singularities, then so does X.

This criterion forms the cornerstone of the proof of the following theorem:

Theorem 15.19. *Let $f : Y \to X$ be a proper surjective morphism with connected fibers between normal varieties. Assume that there exists an effective $\mathbb{Q}$-divisor on Y such that (Y, Δ) is lc and $K_Y + \Delta \sim_{\mathbb{Q},f} 0$. Then X is DB.*

More generally, let $W \subset Y$ be a reduced, closed subscheme that is a union of log canonical centers of (Y, Δ). Then $f(W) \subset X$ is DB.

Corollary 15.20. *Log canonical singularities are DB.*

For the proofs, please see [KK10].

REMARK 15.21. Notice that in (15.18) it is not required that f be birational. On the other hand the assumptions of the theorem and [Kov00b, Thm 1] imply that if $Y \setminus F$ has rational singularities, e.g., if Y is smooth, then $X \setminus W$ has rational singularities as well.

This theorem is used in [KK10] to derive various consequences, some of which are formally unrelated to DB singularities. We will mention some of these in the sequel, but the interested reader should look at the original article to obtain the full picture.

15.B Applications to moduli spaces and vanishing theorems

The connection between log canonical and DB singularities has many useful applications in moduli theory. We will list a few without proof.

SETUP 15.22. Let $\phi : X \to B$ be a flat projective morphism of complex varieties with B connected. Assume that for all $b \in B$ there exists a $\mathbb{Q}$-divisor D_b on X_b such that (X_b, D_b) is log canonical.

REMARK 15.23. Notice that it is not required that the divisors D_b form a family.

Theorem 15.24 [KK10]. *Under the assumptions of* (15.22), *we have the following statements:*

(15.24.1) $h^i(X_b, \mathscr{O}_{X_b})$ *is independent of* $b \in B$ *for all* i.

(15.24.2) If one fiber of ϕ *is Cohen-Macaulay (resp.* S_k *for some* k*), then all fibers are Cohen-Macaulay (resp.* S_k*).*

(15.24.3) The cohomology sheaves $h^i(\omega_\phi^\bullet)$ *are flat over* B, *where* $\omega_\phi^\bullet$ *denotes the relative dualizing complex of* ϕ.

For arbitrary flat, proper morphisms, the set of fibers that are Cohen-Macaulay (resp. S_k) is open, but not necessarily closed. Thus the key point of (15.24.2) is to show that this set is also closed.

The generalization of these results to the semi-log canonical case turns out to be straightforward, but it needs some foundational work to extend some of the results used here to the semi-log canonical case. The general case then implies that each connected component of the moduli space of stable log varieties parameterizes either only Cohen-Macaulay or only non-Cohen-Macaulay objects.

Notice that this still does not mean that we can abandon the non-Cohen-Macaulay objects. There exist smooth projective varieties of general type whose log canonical model is not Cohen-Macaulay and we would naturally prefer to have a moduli space that includes these. Nevertheless, it is very useful to know that if the general fiber is Cohen-Macaulay, then so is the special fiber.

Theorem 15.24 is proved using (15.19) and the following theorem. First we need a simple definition. Let $\phi : X \to B$ be a flat morphism. We say that ϕ is a **DB family** if X_b is DB for all $b \in B$.

Theorem 15.25 [KK10, 7.5]. *Let* $\phi : X \to B$ *be a projective DB family and* $\mathscr{L}$ *a relatively ample line bundle on* X. *Then*

(15.25.1) the sheaves $h^{-i}(\omega_\phi^\bullet)$ *are flat over* B *for all* i,

(15.25.2) the sheaves $\phi_*(h^{-i}(\omega_\phi^\bullet) \otimes \mathscr{L}^{\otimes q})$ *are locally free and compatible with arbitrary base change for all* i *and for all* $q \gg 0$, *and*

(15.25.3) for any base change morphism $\vartheta : T \to B$ *and for all* i,

$$\big(h^{-i}(\omega_\phi^\bullet)\big)_T \simeq h^{-i}(\omega_{\phi_T}^\bullet).$$

DB singularities also appear naturally in vanishing theorems. As a culmination of the work of Tankeev, Ramanujam, Miyaoka, Kawamata, Viehweg, Kollár, and Esnault-Viehweg, Kollár proved a rather general form of a Kodaira-type vanishing theorem in [Kol95, 9.12]. Using the same ideas this was slightly generalized to the following theorem in [KSS10, 6.2].

Theorem 15.26 [Kol95, 9.12], [KSS10, 6.2]. *Let* X *be a proper variety and* $\mathscr{L}$ *a line bundle on* X. *Let* $\mathscr{L}^m \simeq \mathscr{O}_X(D)$, *where* $D = \sum d_i D_i$ *is an effective divisor, and let* s *be a global section whose zero divisor is* D. *Assume that* $0 < d_i < m$ *for every* i. *Let* Z *be the scheme obtained by taking the* m*-th root of* s *(that is,* $Z = X[\sqrt[m]{s}]$ *using the notation from [Kol95, 9.4]). Assume further that*

$$H^j(Z, \mathbb{C}_Z) \to H^j(Z, \mathscr{O}_Z)$$

is surjective. Then for any collection of $b_i \geq 0$ the natural map

$$H^j\left(X, \mathscr{L}^*\left(-\sum b_i D_i\right)\right) \to H^j(X, \mathscr{L}^*)$$

is surjective.

This, combined with the fact that log canonical singularities are DB yields that Kodaira vanishing holds for log canonical pairs:

Theorem 15.27 [KSS10, 6.6]. *Kodaira vanishing holds for Cohen-Macaulay semi-log canonical varieties: Let (X, Δ) be a projective Cohen-Macaulay semi-log canonical pair and $\mathscr{L}$ an ample line bundle on X. Then $H^i(X, \mathscr{L}^*) = 0$ for $i < \dim X$.*

It turns out that DB singularities appear naturally in other kinds of vanishing theorems. Here we mention just one of these.

Theorem 15.28 [GKKP10, 9.3]. *Let (X, D) be a log canonical reduced pair of dimension $n \geq 2$, $\pi : \widetilde{X} \to X$ a log resolution with π-exceptional set E, and $\widetilde{D} = \operatorname{Supp}(E + \pi^{-1}D)$. Then*

$$\mathscr{R}^{n-1}\pi_*\mathscr{O}_{\widetilde{X}}(-\widetilde{D}) = 0.$$

15.C Deformations of DB singularities

Given the importance of DB singularities in moduli theory it is a natural question whether they are invariant under small deformation.

It is relatively easy to see from the construction of the Deligne-Du Bois complex that a general hyperplane section (or more generally, the general member of a base point free linear system) on a variety with DB singularities again has DB singularities. Therefore the question of deformation arises from the following.

Conjecture 15.29 [Ste83]. *Let $D \subset X$ be a reduced Cartier divisor and assume that D has only DB singularities in a neighborhood of a point $x \in D$. Then X has only DB singularities in a neighborhood of the point x.*

This conjecture was proved for isolated Gorenstein singularities by Ishii [Ish86]. Also note that rational singularities satisfy this property, see [Elk78].

We also have the following easy corollary of the results presented earlier:

Theorem 15.30. *Assume that X is Gorenstein and D is normal. Then the statement of Conjecture 15.29 is true.*

Proof. The question is local so we may restrict to a neighborhood of x. If X is Gorenstein, then so is D as it is a Cartier divisor. Then D is log canonical by (15.16), and then the pair (X, D) is also log canonical by inversion of adjunction [Kaw07]. (Recall that if D is normal, then so is X along D). This implies that X is also log canonical and thus DB. □

REMARK 15.31. It is claimed in [Kov00b, 3.2] that the conjecture holds in full generality. Unfortunately, the proof published there is not complete. It works as long as one assumes that the non-DB locus of X is contained in D. For instance, one may assume that this is the case if the non-DB locus is isolated.

The problem with the proof is the following: it is stated that by taking hyperplane sections one may assume that the non-DB locus is isolated. However, this is incorrect. One may only assume that the *intersection* of the non-DB locus of X with D is isolated. If one takes a further general section, then it will miss the intersection point and then it is not possible to make any conclusions about that case.

Therefore currently the best known result with regard to this conjecture is the following:

Theorem 15.32 [Kov00b, 3.2]. *Let $D \subset X$ be a reduced Cartier divisor and assume that D has only DB singularities in a neighborhood of a point $x \in D$ and that $X \setminus D$ has only DB singularities. Then X has only DB singularities in a neighborhood of x.*

Experience shows that divisors not in general position tend to have worse singularities than the ambient space in which they reside. Therefore one would in fact expect that if $X \setminus D$ and D are nice (e.g., they have DB singularities), then perhaps X is even better behaved.

We have also seen that rational singularities are DB and at least Cohen-Macaulay DB singularities are not so far from being rational cf. (15.13). The following result of Schwede supports this philosophical point.

Theorem 15.33 [Sch07, Thm. 5.1]. *Let X be a reduced scheme of finite type over a field of characteristic zero, D a Cartier divisor that has DB singularities and assume that $X \setminus D$ is smooth. Then X has rational singularities (in particular, it is Cohen-Macaulay).*

Let us conclude with a conjectural generalization of this statement:

Conjecture 15.34. *Let X be a reduced scheme of finite type over a field of characteristic zero, D a Cartier divisor that has DB singularities and assume that $X \setminus D$ has rational singularities. Then X has rational singularities (in particular, it is Cohen-Macaulay).*

Essentially the same proof as in (15.30) shows that this is also true under the same additional hypotheses.

Theorem 15.35. *Assume that X is Gorenstein and D is normal. Then the statement of Conjecture 15.34 is true.*

Proof. If X is Gorenstein, then so is D as it is a Cartier divisor. Then by (15.16) D is log canonical. Then by inversion of adjunction [Kaw07] the pair (X, D) is also log canonical near D. (Recall that if D is normal, then so is X along D).

As X is Gorenstein and $X \setminus D$ has rational singularities, it follows that $X \setminus D$ has canonical singularities. Then X has only canonical singularities everywhere. This can be seen by observing that D is a Cartier divisor and examining the discrepancies that lie over D for (X, D) as well as for X. Therefore, by [Elk81], X has only rational singularities along D. $\square$

CHAPTER 16

Subvarieties of moduli spaces

In Chapters 13 and 14 our main focus was the construction of moduli spaces. In this chapter we will review some recent results concerning the structure of these spaces. Moduli theory strives to understand how algebraic varieties deform and degenerate. When studying moduli spaces we are interested in the geometry of the moduli space that reflects the behavior of the families parameterized by the given moduli space. In other words, we are interested in the geometry of **moduli stacks**.

APOLOGY. For the reader uncomfortable with the language of stacks let us promise that we will not use it much. The main point of our choice of using the word "stack" at all is that, as mentioned above, we are interested in the geometry of the moduli space that comes from families. There are subvarieties of a moduli space that do not come from parametrizing a family and here we are not interested in these. When we say "a subvariety of a moduli stack" we really mean the "image of a variety in the moduli stack" or in other words "a subvariety of the moduli space that comes from a parameter space of a family that belongs to the corresponding moduli functor". Of course, to make the material more accessible, we could in fact use this or some similar description to avoid the use of the word "stack". However, we decided not to do that for two main reasons. First of all, the notion of a stack is becoming standard. In a few years, even this paragraph may seem antiquated and completely unnecessary. Second, we **are** talking about subvarieties of a moduli stack when we say so. It seems unwise to try to hide this fact. In addition, our use of stacks is so limited that it requires no knowledge of them at all, only an open mind. We hope that the uninitiated reader will give it a chance and the initiated one will forgive us for not going into more serious detail.

A basic question we are interested in is whether a given moduli stack is proper, or if it is not, then how far it is from being proper. An even simpler question one may ask about the geometry of a given moduli stack is whether it contains any proper subvarieties. And if it does, is there a constraint on what kind of proper subvarieties it may contain?

Naturally, the same questions may be asked about moduli spaces. (As mentioned above, the difference between the two is whether one is interested in any subvariety of the moduli space or only those that come from a family that belongs to the corresponding moduli problem.)

Consider M_g, the moduli space of smooth projective curves of genus $g \geq 3$. M_g admits a projective compactification, called the Satake compactification, with a boundary of codimension 2 [Sat56, BB66, Oor74]. Taking general hyperplane sections on this compactification one finds that M_g contains a proper curve through any point. This does not give explicit families of smooth projective curves that induce a non-constant map from a proper curve to M_g, but at least it shows the existence of such families as we will show next.

Indeed, as the natural map from the moduli stack $\mathscr{M}_g$ to the moduli space M_g is proper, the preimage of a general complete intersection curve through a point is a 1-dimensional substack of $\mathscr{M}_g$. Its normalization is a smooth Deligne-Mumford stack, and so it admits a finite cover by a smooth proper curve. This proves the existence of the image of a proper curve in $\mathscr{M}_g$ through any point. Pulling back the universal family gives the required family over this curve.

If we assume that $g \geq 4$, then the statement that there exists a proper curve through any general point of $\mathscr{M}_g$ is much easier to prove. The locus of curves with non-trivial automorphisms have codimension $g-2$ in M_g. Therefore if $g \geq 4$, then a general complete intersection curve will not intersect that locus and hence it is isomorphic with its preimage via the natural map $\mathscr{M}_g \to \mathsf{M}_g$. This way we obtain a smooth family over the same curve that lives in M_g. In other words, the moduli map of the family is an isomorphism.

EXERCISE 16.1. Let $x_1, \dots, x_r \in \mathsf{M}_g$ be a finite set of points and assume that $g \geq 3$. Prove that there exists a proper curve $C \subset \mathsf{M}_g$ such that $x_1, \dots, x_r \in C$.

EXERCISE 16.2. Use (16.1) to prove that for $g \geq 3$ any global regular function on M_g is constant. In particular, M_g is not affine for $g \geq 3$. Recall, for completeness, that M_g is affine for $g \leq 2$. Prove also that M_g is not projective either (for any $g > 0$). [Hint: use the facts about the Satake compactification mentioned above].

Via a different approach, Kodaira constructed such families explicitly in [Kod67], cf. [Kas68], [BHPV04, V.14], [Zaa95], [GDH99]. On the other hand, this construction is, in some sense, the exact opposite of the one above; the images of these curves in the corresponding moduli stack $\mathscr{M}_g$ (or in the moduli space M_g) are confined to the special locus of curves that admit non-trivial automorphisms, so they never contain a general point of the moduli space/stack.

These results naturally lead to the following question: Are there higher dimensional proper subvarieties contained in some $\mathscr{M}_g$? The answer is affirmative. Kodaira's construction or the Satake compactification may be used to prove the following: For any $d \in \mathbb{N}$ there exists a $g = g(d) \in \mathbb{N}$ such that $\mathscr{M}_g$ contains a proper subvariety of dimension d. For details on this construction see [Mil86], [FL99, pp.34-35], [Zaa99]. These examples are all based on the aforementioned construction of Kodaira and hence the proper subvarieties constructed this way all lie in the locus of curves that admit a morphism onto another curve of positive genus.

One may argue that the really interesting question is whether there are higher dimensional proper subvarieties of $\mathscr{M}_g$ that contain a general point of $\mathscr{M}_g$ cf. [Oor95]. By the above argument, we know that this is true for curves, but unfortunately, this is still an open question even for surfaces, i.e., it is not known whether there are proper surfaces through a general point of $\mathscr{M}_g$, for any $g > 3$.

Naturally, since $\dim \mathscr{M}_g = 3g - 3$, there is an obvious upper bound on the dimension of a proper subvariety of $\mathscr{M}_g$ for a fixed g, but one may ask whether there is a significantly better upper bound than $3g-3$. Actually this is one of those questions when finding the answer for the moduli space, M_g, implies the same for the moduli stack, $\mathscr{M}_g$, and not the other way around. The celebrated theorem of Diaz-Looijenga [Dia84, Dia87, Loo95] says that any proper subvariety of M_g has dimension at most $g - 2$. This estimate is trivially sharp for $g = 2$ and in view of Kodaira's construction mentioned above it is also sharp for $g = 3$, but it is not known to be sharp for any other values of g. The known examples are very far from this bound. Recently, Faber and van der Geer [FvdG04] pointed out that in char p there exists a natural subvariety of $\mathscr{M}_g$ of expected dimension $g - 2$, and hence seems a good candidate for (containing) a proper subvariety of maximal dimension. However, they also show that this subvariety has some non-proper components and hence itself is not proper. On the other hand, Faber and van der Geer express hope that it might also have proper components. This would be enough to prove that the upper bound $g - 2$ is sharp.

Similar questions may be asked about other moduli spaces/stacks, for instance, replacing curves by abelian varieties. The reader interested in this question could start by consulting [Oor74], [KS03], and [VZ05c].

One of the topics we are discussing in this chapter is a somewhat more sophisticated question. On one hand, we are not only asking whether a given moduli stack contains proper subvarieties, but we would like to know what kind of proper subvarieties it contains. For instance, does it contain proper rational or elliptic curves? We can even take it a step further and ask: If it does not contain a proper rational curve, does it contain a curve that's isomorphic to the affine line?

Interestingly, already the question of containing proper rational curves differentiates between the moduli stack, $\mathscr{M}_g$, and the moduli space, M_g: Parshin [Par68] proved that $\mathscr{M}_g$ does not contain proper rational curves for any g, while Oort [Oor74] showed that there exists some g such that M_g does contain proper rational curves.

Our starting point in this chapter is Shafarevich's conjecture (16.4). This leads us to investigate related questions and eventually to a recent generalization, Viehweg's conjecture (16.45), which states that any subvariety of the moduli stack is of log general type.

This topic has gone through an enormous transformation during the last decade and consequently it is impossible to cover all the new developments in as much detail as they deserve. Hence the reader is encouraged to consult other related surveys [Vie01], [Kov03a], [MVZ05], [Kov09].

16.A Shafarevich's conjecture

Let us start with the aforementioned conjecture of Shafarevich [Sha63]. Recall the definition of an *admissible* family from (2.17):

DEFINITION 16.3. Let B be a smooth variety over k, and $\Delta \subseteq B$ a closed subset. Further let $h \in \mathbb{Q}[t]$ be a polynomial. A family $f: X \to B$ is **admissible** (with respect to (B, Δ) and h) if

(16.3.1) X is smooth,

(16.3.2) $f: X \to B$ is not isotrivial,

(16.3.3) $f: X \setminus f^{-1}(\Delta) \to B \setminus \Delta$ is smooth, and

(16.3.4) X_b is projective, ω_{X_b} is ample with Hilbert polynomial $h(m) = \chi(X, \omega_{X_b}^{\otimes m})$ for all $b \in B \setminus \Delta$.

Two admissible families are **equivalent** if they are isomorphic over $B \setminus \Delta$.

16.4 Shafarevich's conjecture. *Let (B, Δ) be fixed and $q \geq 2$ an integer. Then:*

(16.4.1) There exist only finitely many isomorphism classes of admissible families of curves of genus q.

(16.4.2) If $2g(B) - 2 + \#\Delta \leq 0$, then there exist no such families.

REMARK 16.5. Note that this statement is usually known as the *function field case* of Shafarevich's conjecture. In this chapter we will mainly concentrate on this case, so we will omit the qualifier and refer to it simply as *Shafarevich's conjecture* cf. (16.11).

EXERCISE 16.6. Prove that the inequality in (16.4.2) can be satisfied only if B is either a rational or an elliptic curve:

$$2g(B) - 2 + \#\Delta \leq 0 \quad \Leftrightarrow \quad \begin{cases} g(B) = 0 & \text{and} \quad \#\Delta \leq 2, \\ g(B) = 1 & \text{and} \quad \Delta = \emptyset. \end{cases}$$

Shafarevich showed a special case of (16.4.2): There exist no smooth families of curves of genus q over $\mathbb{P}^1$. (16.4.1) was confirmed by Parshin [Par68] for $\Delta = \emptyset$ and by Arakelov [Ara71] in general.

Our main goal is to generalize this statement to higher dimensional families. In order to do this we will have to reformulate the statement as Parshin and Arakelov did.

16.B The Parshin-Arakelov reformulation

With regard to Shafarevich's conjecture, Parshin made the following observation. In order to prove that there are only finitely many admissible families, one can try to proceed the following way. Instead of aiming for the general statement, first try to prove that there are only finitely many deformation types (2.G). The next step then is to prove that admissible families are rigid, that is, they do not admit non-trivial deformations. Notice that if we prove these statements for families over $B \setminus \Delta$, then they also follow for families over B. Now since every deformation type contains only one family, and since there are only finitely many deformation types, the original statement follows.

The following is the reformulation of Shafarevich's conjecture that was used by Parshin and Arakelov:

16.7 Shafarevich's conjecture (version two). *Let (B, Δ) be fixed and $q \geq 2$ an integer. Then the following statements hold.*

- **(B)** (BOUNDEDNESS) *There exist only finitely many deformation types of admissible families of curves of genus q with respect to $B \setminus \Delta$.*
- **(R)** (RIGIDITY) *There exist no non-trivial deformations of admissible families of curves of genus q with respect to $B \setminus \Delta$.*
- **(H)** (HYPERBOLICITY) *If $2g(B) - 2 + \#\Delta \leq 0$, then no admissible families of curves of genus q exist with respect to $B \setminus \Delta$.*

REMARK 16.8. As we discussed above, **(B)** and **(R)** together imply (16.4.1) and **(H)** is clearly equivalent to (16.4.2).

16.C Shafarevich's conjecture for number fields

The number field version of Shafarevich's conjecture played a prominent role in Faltings' proof of the Mordell conjecture. This section is a brief detour to this very exciting area, but it is disconnected from the rest of the chapter. The reader should feel free to skip this section and continue with the next one.

DEFINITION 16.9. Let $(R, \mathfrak{m})$ be a DVR, $F = \operatorname{Frac}(R)$, and C a smooth projective curve over F. C is said to have **good reduction over** R if there exists a scheme Z, smooth and projective over Spec R, such that $C \simeq Z_F$,

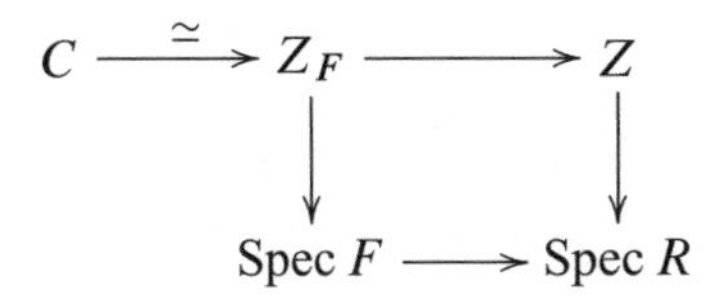

DEFINITION 16.10. Let R be a Dedekind ring, $F = \operatorname{Frac}(R)$, and C a smooth projective curve over F. C has **good reduction at the closed point** $\mathfrak{m} \in \operatorname{Spec} R$ if it has good reduction over $R_{\mathfrak{m}}$.

16.11 Shafarevich's conjecture (number field case). *Let $q \geq 2$ be an integer.*

(16.11.1) Let F be a number field, $R \subset F$ the ring of integers of F, and $\Delta \subset \operatorname{Spec} R$ a finite set. Then there exists only finitely many smooth projective curves over F of genus q that have good reduction outside Δ.

(16.11.2) There are no smooth projective curves of genus q over Spec $\mathbb{Z}$.

REMARK 16.12. Shafarevich's conjecture in the number field case has been confirmed: (16.11.1) by Faltings [Fal83b, Fal84] and (16.11.2) by Fontaine [Fon85].

One can reformulate (16.4.1) to resemble the above statement. Note that a curve over a funcion field $F = K(B)$ is the same as the general fiber of a family of curves over B. It is **non-isotrivial** if the corresponding family is non-isotrivial.

16.13 Shafarevich's conjecture (version three). *Let $q \geq 2$ be an integer, B a smooth complex projective curve and $F = K(B)$ the function field of B. Let $\Delta \subset B$ be a finite subset such that $B \setminus \Delta = \operatorname{Spec} R$ for a (Dedekind) ring R. Then there exist only finitely many smooth projective non-isotrivial curves of genus q over F having good reduction over all closed points of* Spec R.

DEFINITION 16.14. If C is a smooth projective curve over F (an arbitrary field), then there exists a morphism $C \to \operatorname{Spec} F$. Sections, $\operatorname{Spec} F \to C$, of this morphism correspond in a one-to-one manner to **F-rational points** of C, points that are defined over the field F. F-rational points of C will be denoted by $C(F)$.

EXAMPLE 16.15. The $\mathbb{R}$-rational points of the complex projective curve $x^2 + y^2 - z^2 = 0$ form a circle, its $\mathbb{C}$-rational points form a sphere.

EXAMPLE 16.16. The projective curve $x^2 + y^2 + z^2 = 0$ has no $\mathbb{R}$-rational points.

EXAMPLE 16.17. Let C_n be the (complex) projective curve defined by the equation $x^n + y^n - z^n = 0$. It follows from Wiles' theorem (Fermat's last theorem) [Wil95a, Wil95b], that for $n \geq 3$,

$$C_n(\mathbb{Q}) = \begin{cases} \{[1:0:1],[0:1:1],[1:-1:0]\}, & \text{if } n \text{ is odd,} \\ \{[1:0:1],[0:1:1],[1:0:-1],[0:1:-1]\} & \text{if } n \text{ is even.} \end{cases}$$

As mentioned earlier, Faltings [Fal83b, Fal84] used (16.11) to prove:

16.18 Faltings' theorem (Mordell's conjecture). *Let F be a number field and C a smooth projective curve of genus $q \geq 2$ defined over F. Then $C(F)$ is finite.*

The function field version of this conjecture was proved earlier by Manin [Man63]:

16.19 Manin's theorem (Mordell's conjecture for function fields). *Let F be a function field (i.e., the function field of a variety over k, where k is an algebraically closed field of characteristic* 0*) and let C be a smooth projective non-isotrivial curve over F of genus $q \geq 2$. Then $C(F)$ is finite.*

REMARK 16.20. The essential case to settle is when $\operatorname{tr.deg}_k F = 1$, i.e., $F = K(B)$, where B is a smooth projective curve over k.

16.D From Shafarevich to Mordell: Parshin's trick

Shafarevich's conjecture implies Mordell's in both the function field and the number field case by an argument due to Parshin. The first step is a clever way to associate different (families of) curves to different sections:

16.21 Parshin's covering trick. *For every F-rational point, $P \in C(F)$, or equivalently, for every section $X \overset{\sigma_P}{\curvearrowleft\!\!\to} B$, there exists a finite cover of X, $W_P \xrightarrow{\pi_P} X$ such that*

- *the degree of π_P is bounded in terms of q,*
- *the projection $W_P \to B$ is smooth over $B \setminus \Delta$,*

- *the map π_P is ramified exactly over the image of σ_P,*
- *the genus of the fibers of $W_P \to B$ is bounded in terms of q.*

For details on the construction of the covers, $W_P \xrightarrow{\pi_P} X$, see [Lan97, IV.2.1] and [Cap02, §4]. The second step is an old result:

16.22 de Franchis's theorem [dF13, dF91]. *Let C and D be two fixed smooth projective curves of genus at least 2. Then there exist only finitely many dominant rational maps $D \to C$.*

Shafarevich's conjecture implies that there are only finitely many different W_P's. Viewing W_P and X as curves over F, de Franchis's theorem implies that a fixed W_P can admit only finitely many different maps to X.

Since those maps are ramified exactly over the image of the corresponding σ_P, this means that there are only finitely many σ_P's, i.e., $C(F)$ is finite, and therefore Mordell's conjecture follows from that of Shafarevich.

We end our little excursion to the number field case here. In the rest of the chapter we work in the function field case and use the original notation and assumptions.

16.E Hyperbolicity and Boundedness

16.E.1 Hyperbolicity

DEFINITION 16.23. [Bro78] A complex analytic space X is called **Brody hyperbolic** if every holomorphic map $\mathbb{C} \to X$ is constant.

REMARK 16.24. Another important, related analytic notion is Kobayashi hyperbolicity. For its definition the reader is referred to [Kob70] and [Lan87]. Let us just mention here that for a compact analytic space X, being Brody hyperbolic is equivalent to being Kobayashi hyperbolic. In general, if X is not compact, then Kobayashi hyperbolicity implies Brody hyperbolicity but not vice versa.

EXERCISE 16.25. Prove that a complex analytic space X is Brody hyperbolic if and only if every holomorphic map $\mathbb{C}^\times \to X$ is constant.

EXERCISE 16.26. Let T be a complex torus and X a Brody hyperbolic complex analytic space. Prove that then every holomorphic map $T \to X$ is constant.

We would like to define the algebraic analog of hyperbolicity motivated by this observation. Algebraic maps are more restrictive than holomorphic ones. For instance the universal covering map, $\mathbb{C} \to E$, of an elliptic curve, E, is not algebraic. In particular, excluding algebraic maps from $\mathbb{C}$ to X does not exclude maps from E to X as in the analytic case (cf. (16.26)). The same argument goes for abelian varieties. Since there exist simple abelian varieties (those that do not contain other abelian varieties) of arbitrary dimension, we have to take into consideration arbitrary dimensional abelian varieties.

The following definition of *algebraic hyperbolicity* is an algebraic version of Brody hyperbolicity and perhaps it should be called "algebraic Brody hyperbolicity" to emphasize that fact. However, this is not the established terminology.

A further complication is that there are some related, but different definitions that are used with the same name [Dem97], [Che04]. As usual in similar cases, these different variants were introduced around the same time and developed parallel to each other. Hence it is hard to go back and change the terminology. We will also introduce the notion of **weak boundedness**, which is closer in spirit to Demailly's notion of hyperbolicity. As the reader will see, the known results of hyperbolicity (as used in this chapter) follow from weak boundedness (16.31), hence the statements actually remain true even if one uses Demailly's definition of algebraic hyperbolicity.

One major advantage of the definition used here is that it extends naturally to stacks, which is exactly the context in which we would like to use it.

DEFINITION 16.27. An algebraic stack $\mathscr{M}$ is called **algebraically hyperbolic** if

– every morphism $\mathbb{A}^1 \setminus \{0\} \to \mathscr{M}$ is constant, and

– every morphism $A \to \mathscr{M}$ is constant for any abelian variety, A.

REMARK 16.28. The first row in the following diagram is the statement of condition (**H**). The last row shows equivalent conditions for both the assumption and the conclusion. Recall that $\mathscr{M}_q$ stands for the moduli stack (of curves of genus q), so maps of the form $B \setminus \Delta \to \mathscr{M}_q$ are exactly the ones that are induced by families over $B \setminus \Delta$.

$$\begin{array}{ccc} 2g(B) - 2 + \#\Delta \leq 0 & \Longrightarrow & \nexists\, f\colon X \to B \text{ admissible} \\ \Updownarrow & & \Updownarrow \\ B \setminus \Delta \supseteq \mathbb{A}^1 \setminus \{0\} \text{ or,} & & \nexists\, B \setminus \Delta \to \mathscr{M}_q \\ \Delta = \emptyset \text{ and } B \text{ is an elliptic curve} & & \text{non-constant} \end{array}$$

This implies that proving (**H**) is equivalent to proving that there does not exist a non-constant morphism of the form $\mathbb{A}^1 \setminus \{0\} \to \mathscr{M}_q$ or $E \to \mathscr{M}_q$, where E is an arbitrary elliptic curve.

Corollary 16.29. *If $\mathscr{M}_q$ is algebraically hyperbolic, then* (**H**) *holds.*

16.E.2 Weak Boundedness

In addition to properties (**B**), (**R**), and (**H**), there is another important property to study. Its importance lies in the fact that it implies (**H**) and if some additional conditions hold it also implies (**B**).

(**WB**) (WEAK BOUNDEDNESS) *For an admissible family $f\colon X \to B$, the degree of $f_*\omega_{X/B}^{\otimes m}$ is bounded above in terms of $g(B), \#\Delta, g(X_{\mathrm{gen}}), m$. In particular, the bound is independent of f and B.*

The traditional proof of hyperbolicity for curves proceeds via some form of weak boundedness. The key point is that the upper bound obtained for $\deg f_*\omega_{X/B}^{\otimes m}$ has the form

$$(2g(B) - 2 + \#\Delta) \cdot c(g(B), \#\Delta, g(X_{\mathrm{gen}}), m),$$

where $c(g(B), \#\Delta, g(X_{\mathrm{gen}}), m)$ is a positive constant depending only on the indicated values. This proves hyperbolicity: Since $\det f_*\omega_{X/B}^{\otimes m}$ is ample, its degree is positive, so any upper bound of it is also positive.

In higher dimensions, the bounds obtained are not always of this form. However, perhaps somewhat surprisingly, hyperbolicity follows already from the fact of weak boundedness, not only from the explicit bound.

Theorem 16.30. (WB) $\Rightarrow$ (H).

A more precise and somewhat more general formulation is the following:

Theorem 16.31 [Kov02, 0.9] cf. [Par68]. *Let $\mathfrak{F}$ be a collection of smooth varieties of general type, B a smooth projective curve and $\Delta \subset B$ a finite subset of B. Let*

$$\mathrm{Fam}(B, \Delta, \mathfrak{F}) = \{f : X \to B \mid X \text{ is smooth, } f \text{ is flat and } f^{-1}(t) \in \mathfrak{F} \text{ for all } t \in B \setminus \Delta\}.$$

Assume that $\mathrm{Fam}(B, \Delta, \mathfrak{F})$ *contains non-isotrivial families and that there exist $M, m \in \mathbb{N}$ such that for all $(f : X \to B) \in \mathrm{Fam}(B, \Delta, \mathfrak{F})$,*

$$\deg\left(f_*\omega_{X/B}^{\otimes m}\right) \leq M.$$

Then $2g(B) - 2 + \#\Delta > 0$.

Proof. Assume the contrary, i.e., either $g(B) = 0$ and $\#\Delta \leq 2$ or $g(B) = 1$ and $\#\Delta = 0$. This allows us to assume that $f : X \to B$ is semi-stable and non-isotrivial. Also, in both cases there exists a finite endomorphism, $\tau : B \to B$, of degree > 1 such that τ is smooth over $B \setminus \Delta$ and completely ramified over Δ.

Let $\pi : \tilde{X}_\tau \to X_\tau$ be a resolution of singularities that is an isomorphism over $B \setminus \Delta$, and $\tilde{f}_\tau = f_\tau \circ \pi$. Then $(\tilde{f}_\tau : \tilde{X}_\tau \to B_\tau) \in \mathrm{Fam}(B, \Delta, \mathfrak{F})$. In particular $B_\tau \simeq B$ and f_τ is smooth over $B_\tau \setminus \Delta_\tau \simeq B \setminus \Delta$ with fibers in $\mathfrak{F}$.

Therefore, by assumption $\deg\tau \cdot \deg(f_*\omega_{X/B}^m) = \deg(f_{\tau *}\omega_{X_\tau/B_\tau}^m) \leq M$ as well. On the other hand, $\deg(f_*\omega_{X/B}^m) > 0$ by [Kol87a], which then implies that $\deg\tau$ is bounded. However, by iterating the above process, $\deg\tau$ can grow arbitrary large, so we obtain a contradiction and the statement is proved. □

16.E.3 From Weak Boundedness to Boundedness

By [Par68, Theorem 1] there exists a scheme V that parameterizes admissible families of curves of genus q. Hence **(B)** is equivalent to the statement that V has finitely many components. Therefore, in order to prove **(B)**, it is enough to prove that V is a scheme of finite type.

V maps naturally to $\mathrm{Hom}((B, B \setminus \Delta), (\overline{\mathsf{M}}_q, \mathsf{M}_q)) \subset \mathrm{Hilb}(B \times \overline{\mathsf{M}}_q)$. For a family $f : X \to B$, let $\mu_f : B \to \overline{\mathsf{M}}_q$ be the moduli map. If for a fixed ample line bundle $\mathscr{L}$ on $\overline{\mathsf{M}}_q$, one can establish that $\deg \mu_f^*\mathscr{L}$ is bounded on B, the bound perhaps depending on B, Δ and q, but not on f, then one may conclude that the image of V in $\mathrm{Hilb}(B \times \overline{\mathsf{M}}_q)$ is contained in finitely many components and hence is of finite type.

The final piece of the puzzle is provided by the construction of $\overline{\mathsf{M}}_q$. For p sufficiently large and divisible there exist line bundles $\lambda_m^{(p)}$ on $\overline{\mathsf{M}}_q$ such that for a family of stable curves $f: X \to B$, if $\bar{\mu}_f: B \to \overline{\mathsf{M}}_q$ is the induced moduli map, then

$$\left(\det\left(f_*\omega_{X/B}^{\otimes m}\right)\right)^p = \bar{\mu}_f^* \lambda_m^{(p)}.$$

Hence (**WB**) gives exactly the above required boundedness result and so we obtain the following statement.

Theorem 16.32. *For families of curves,* (**WB**) *implies* (**B**).

REMARK 16.33. The reader might be wondering why we did not use stacks here. The answer is simple. Using stacks would be the natural thing, but we do not really need them here and we have promised to minimize the use of stacks.

The key point of this step is that weak boundedness is equivalent to the statement that the pull-back of a natural line bundle from the moduli *space* has bounded degree. The stacky part here is actually contained in the existence of V. Although Parshin did not use stacks to prove this, that's where it would be natural to use them. In fact, the similar statement in higher dimensions had been unknown until very recently. Kovács and Lieblich established the existence of the analogue of V for admissible families of higher dimensional varieties in [KL06]. For more on this see (16.F.4) and (16.40).

16.F Higher dimensional fibers

Next we turn to higher dimensional generalizations. First, we will keep the assumption that the base of the family is a curve, but allow the fibers to have higher dimensions. Independently, or simultaneously, one can study families over higher dimensional bases, and we will do that in the next section. Furthermore, generalizing the conditions on the fibers naturally leads to the study of families of singular varieties. We will discuss all of these directions.

In order to generalize Shafarevich's conjecture to the case of families of higher dimensional varieties, the first task is to generalize both the statement and the conditions. The condition that a curve has genus at least 2, i.e., our assumption that $g(X_{\text{gen}}) \geq 2$, is equivalent to the condition that $\omega_{X_{\text{gen}}}$ is ample. In higher dimensions, the role of the genus is played by the Hilbert polynomial, so fixing $g(X_{\text{gen}})$ will be replaced by fixing $h_{\omega_{X_{\text{gen}}}}$, the Hilbert polynomial of $\omega_{X_{\text{gen}}}$. Therefore we have the following starting data:

- a fixed smooth curve B of genus $g = g(B)$,
- a fixed finite subset $\Delta \subset B$, and
- a fixed polynomial h.

We have made the definition of an admissible family in (2.17) so the various parts of Shafarevich's conjecture make sense in any dimension.

16.34 Higher dimensional Shafarevich conjecture. *Fix B, Δ and h. Then*

- **(B)** (BOUNDEDNESS) *there exist only finitely many deformation types of admissible families of canonically polarized varieties with respect to B, Δ and h,*
- **(R)** (RIGIDITY) *there exist no non-trivial deformations of admissible families of canonically polarized varieties with respect to B, Δ and h,*
- **(H)** (HYPERBOLICITY) *if $2g(B) - 2 + \#\Delta \leq 0$, then no admissible families of canonically polarized varieties with respect to B, Δ and h exist, and*
- **(WB)** (WEAK BOUNDEDNESS) *for an admissible family $f: X \to B$, the degree of $f_* \omega_{X/B}^{\otimes m}$ is bounded in terms of $g(B), \#\Delta, h$ and m.*

Next we will discuss the state of affairs with regard to these conjectures and the many results obtained during the past decade. Because of the interdependency of the various results it makes more sense to follow a different order than they are listed in the conjecture.

16.F.1 Rigidity

Let $Y \to B$ be an arbitrary non-isotrivial family of curves of genus ≥ 2, and C a smooth projective curve also of genus ≥ 2. Then $f: X = Y \times C \to B$ is an admissible family, and a deformation of C gives a deformation of f. Therefore **(R)** fails as stated.

This leads naturally to the following question.

Question 16.35. Under what additional conditions does **(R)** hold?

A possible answer to this question will be given in (16.K).

16.F.2 Hyperbolicity

Migliorini [Mig95] showed that for families of minimal surfaces a somewhat weakened hyperbolicity statement holds, namely that $\Delta \neq \emptyset$ if $g \leq 1$. The same conclusion was shown in [Kov96] for families of minimal varieties of arbitrary dimension. Later **(H)** for families of minimal surfaces was proved in [Kov97b], and then in general for families of canonically polarized varieties in [Kov00a].

Theorem 16.36 [Kov00a]. *Let $X \to B$ be an admissible family of canonically polarized varieties with respect to B, Δ and h. Then $2g(B) - 2 + \#\Delta > 0$.*

Finally, Viehweg and Zuo [VZ03b] proved the analytic version of **(H)**:

Theorem 16.37 [VZ03b]. *$\mathcal{M}_h$ is Brody hyperbolic.*

16.F.3 Weak Boundedness

Bedulev and Viehweg [BV00] proved the following:

Theorem 16.38 [BV00]. *Let $f: X \to B$ be an admissible family with B, Δ, h fixed. Let $\delta = \#\Delta$, $g = g(B)$, and $n = \dim X_{\text{gen}} = \dim X - 1$. If $f_* \omega_{X/B}^{\otimes m} \neq 0$, then there exists a positive integer $e = e(m, h)$ such that*

$$\deg f_* \omega_{X/B}^{\otimes m} \leq m \cdot e \cdot \operatorname{rk} f_* \omega_{X/B}^{\otimes m} \cdot (n(2g - 2 + \delta) + \delta).$$

This clearly implies (**WB**) and as a byproduct of the explicit bound it also implies (**H**).

Viehweg and Zuo [VZ01] extended (**WB**) to families of varieties of general type and of varieties admitting a minimal model with a semi-ample canonical bundle. In [Kov02] similar results were obtained with different methods allowing the fibers to have rational Gorenstein singularities, but restricting to the case of families of minimal varieties of general type.

The proof of (16.31) still works in this generality, so (**WB**) implies (**H**) in all dimensions [Kov02, 0.9].

16.F.4 Boundedness

Using the existence of moduli spaces of canonically polarized varieties and the description of ample line bundles on them, Bedulev and Viehweg [BV00] also proved a boundedness-type statement:

Theorem 16.39 [BV00]. *Let B, Δ and h be fixed and assume that M_h admits a modular compactification $\overline{\mathsf{M}}_h$. Then there exists a subscheme of* $\mathrm{Hom}((B, B \setminus \Delta), (\overline{\mathsf{M}}_h, \mathsf{M}_h))$ *of finite type that contains the classes of all morphisms $B \to \overline{\mathsf{M}}_h$ induced by admissible families.*

Unfortunately this statement does not imply (**B**). However, a recent result of Kovács and Lieblich does.

Theorem 16.40 [KL06]. *Let B, Δ and h be fixed. Then there exist only finitely many deformation types of admissible families of canonically polarized varieties with respect to B, Δ and h.*

16.F.5 Shafarevich's conjecture for other types of varieties

One may ask if there is a positive answer to Shafarevich's problem for families of other types of varieties. There are some known results in this setting as well.

Faltings [Fal83a] studied the Shafarevich problem for families of abelian varieties and proved that (**B**) holds, while (**R**) fails in general. He also gave an equivalent condition for (**R**) to hold in this case.

Oguiso and Viehweg [OV01] considered (**H**) for families of non-general type surfaces. Their work combined with the previous results show that (**H**) holds for families of minimal surfaces of non-negative Kodaira dimension.

Recent results have been obtained by Jorgensen and Todorov [JT02], Liu, Todorov, Yau and Zuo [LTYZ05] and Viehweg and Zuo [VZ05b] for families of Calabi-Yau varieties.

16.G Higher dimensional bases

The next natural generalization is to allow B to have arbitrary dimension. Let B be a smooth projective variety, $\Delta \subset B$ a divisor with normal crossings and h a polynomial.

The definition of an admissible family is the same as before (2.17). As before, for an admissible family, $f: X \to B$, the moduli map $b \mapsto [X_b]$ is denoted by $\mu_f: B \setminus \Delta \to \mathscr{M}_h$.

Since B is now allowed to be higher dimensional, the notion of isotriviality is no longer the best one to consider. Observe that f is isotrivial if and only if μ_f is constant. Saying that f is not isotrivial would allow the family to be isotrivial in certain directions. What we want to assume is that the family "truly" changes in any direction on B. To express this we define the family's variation in moduli.

DEFINITION 16.41 [Vie83a, Vie83b, Kol87b]. $\operatorname{Var} f := \dim(\mu_f(B))$ $(\leq \dim B)$.

We are interested in the case $\operatorname{Var} f = \dim B$. In (16.29), we observed that hyperbolicity follows if we know that the stack $\mathscr{M}_h$ is algebraically hyperbolic. In fact, for hyperbolicity over a 1-dimensional base, we only needed the corresponding property of $\mathscr{M}_h$ for curves. However, we would also like to know that every morphism $A \to \mathscr{M}_h$ induced by a family is constant, where A is an arbitrary abelian variety. This is the extra content of the next theorem.

Theorem 16.42 [Kov97a, Kov00a]. *$\mathscr{M}_h$ is algebraically hyperbolic.*

As before, this implies that if $f: X \to \mathbb{P}^1$ is an admissible family, then $\#\Delta > 2$ (16.31). More generally, for an admissible family $f: X \to \mathbb{P}^m$ with $\operatorname{Var} f = m$, this implies that $\deg \Delta > 2$. However, we expect that in this case $\deg \Delta$ should be larger than $m + 1$. This is indeed the case.

Theorem 16.43 [VZ02, Kov03c]. *Let $f: X \to \mathbb{P}^m$ be an admissible family. Then $\omega_{\mathbb{P}^m}(\Delta)$ is ample, or equivalently $\deg \Delta > m + 1$.*

REMARK 16.44. Viehweg and Zuo actually prove a lot more than this in [VZ02]. Please see the article for details.

It is now natural to suspect that a more general statement should hold. The following statement to this effect is part of a more general conjecture of Viehweg [Vie01].

16.45 Viehweg's conjecture. *If $f: X \to B$ is an admissible family, then $\omega_B(\Delta)$ is big.*

For $\dim B = 1$, this is simply **(H)**. For $\dim B > 1$, it is known to be true for families of curves by [Vie01, 2.6] and for $B = \mathbb{P}^n$ and various other special cases by [VZ02], [Kov03b] and [Kov03c]. It was confirmed for $\dim B = 2$ by Kebekus and Kovács in [KK08a]. In fact, in this latter paper (16.45) was refined and the following more general conjecture was confirmed for the case of $\dim B = 2$:

16.46 Refined Viehweg's conjecture [KK08a, 1.6]. *Let $f : X \to B$ be an admissible family. Then either $\kappa(B, \Delta) = -\infty$ and $\operatorname{Var}(f) < \dim B$, or $\operatorname{Var}(f) \leq \kappa(B, \Delta)$.*

Assuming that the minimal model program (1.28), including the abundance conjecture, works for the base, (16.46) was recently proved for the case when B is compact (of arbitrary dimension), i.e., that $\Delta = \emptyset$, by Kebekus and Kovács in [KK08b]. In particular, it follows that (the refined) Viehweg's conjecture is true for compact bases up to dimension 3. The recent results explained in Part II of this book show that some of these assumptions are always satisfied. The only currently missing piece is the abundance conjecture, especially the consequence that a smooth projective variety with negative Kodaira dimension is necessarily uniruled.

The methods used in [KK08a] were limited to $\dim B = 2$ for various reasons without much hope to extend the results to higher dimensions. A new approach was taken in [KK07] that led to a different way of attacking Viehweg's conjecture through a seemingly unrelated problem of extending logarithmic differential forms over singularities. New results on this extension problem were recently obtained by Greb, Kebekus and Kovács in [GKK10] and these extension results were used by Kebekus and Kovács in [KK08c] to settle the refined Viehweg's conjecture for arbitrary B of dimension at most 3. Furthermore, [KK08c] contains a description of the induced moduli map via Kodaira and Mori fibrations. Please see that paper for details.

Despite the many recent advances, this question is still far from being completely settled. The reader is encouraged to read Viehweg's discussion of this and other related open questions in [Vie01].

16.H Uniform and effective bounds

16.H.1 Families of curves

A finiteness result such as (16.4.1) naturally leads to the question of whether the obtained bound depends on the actual curve, or only on its genus. In other words, is it possible to give a **uniform bound** that works for all base curves B of genus g?

This question is actually more subtle than it might seem at first. Consider the argument preceding (16.32). That argument proves that (**WB**) implies (**B**), but it does not shed any light on the obtained bound. Even if the bound appearing in (**WB**) depends only on the genus, it might happen that the number of deformation types still depends on the actual curve. The argument uses the fact that a subscheme of a scheme of finite type itself is of finite type. This means that the subscheme has finitely many components, which is what is needed for (**B**), but it says nothing about how big that finite number is. The number of components of a subscheme has nothing to do with the number of components of the ambient scheme.

Despite these difficulties, uniform boundedness is known. The first such result was obtained by Caporaso:

Theorem 16.47 [Cap02, Cap03, Cap04]. *There exists a constant $c(q,d,\delta)$ such that for any smooth irreducible variety $B \subseteq \mathbb{P}^r$ of* degree d *and for any closed subscheme $\Delta \subset B$ of degree δ, the number of admissible families of curves of genus q with respect to (B,Δ) is at most $c(q,d,\delta)$.*

REMARK 16.48. If B is one-dimensional, then one may write $c(q,d,\delta) = c'(q,g,\delta)$ using $g = g(B)$ the genus of B instead of d.

The next question is whether the constant $c(q,d,\delta)$ (or in the case of a base curve $c'(q,g,\delta)$) is computable. In other words, is it possible to give an *effective* uniform bound? For families over a base curve this was achieved by Heier:

Theorem 16.49 [Hei04]. *Let B be a smooth projective curve of genus g and $\Delta \subset B$ a finite subset. Then $c'(q, g, \delta)$ can be expressed as an explicit function of q, g and δ.*

REMARK 16.50. The expression itself is rather complicated and can be found in the original article.

16.H.2 Higher dimensional families

For higher dimensional families rigidity fails and so we cannot expect a similar finiteness statement as above. However, one may still ask whether **uniform boundedness** holds and if so, whether there is an effective bound on the number of deformation types over a base with a fixed Hilbert polynomial.

Uniform boundedness was proved by Kovács and Lieblich.

Theorem 16.51 [KL06]. *Let h be a fixed polynomial. Then the set of deformation types of admissible families of canonically polarized varieties with Hilbert polynomial h is finite and uniformly bounded in any quasi-compact family of base varieties B°.*

Recently Heier proved a uniformity statement where the bound depends on the dimension and the top self-intersection number of the canonical class of the fibers and the total space.

Theorem 16.52 [Hei09]. *The number of deformation types of admissible families $f : X \to B$ of canonically polarized varieties with a fixed Hilbert polynomial and fixed numerical values K_X^{n+1} and K_F^n, where F is the general fiber of f and $n = \dim F$, is no more than an explicitly computable number that depends only on the values n, K_X^{n+1}, K_F^n, $g(B)$.*

For the actual value of the bound please see the original article. Also note that it would be interesting to see a uniformity statement where one does not have to restrict K_X^{n+1}.

16.I Techniques

16.I.1 Positivity of direct images

One of the most important ingredients in the proofs of the known results is an appropriate variant of a fundamental positivity result due to the work of Fujita, Kawamata, Kollár and Viehweg. In this section we will assume, for simplicity, that $\dim B = 1$.

DEFINITION 16.53. A locally free sheaf, $\mathscr{E}$, is **ample** if $\mathscr{O}_{\mathbb{P}(\mathscr{E})}(1)$ on $\mathbb{P}(\mathscr{E})$ is ample.

Theorem 16.54 [Fuj78, Kaw82b, Vie83a, Vie83b, Kol87b, Kol90a]. *Let $f : X \to B$ be an admissible family and $m > 1$. If $f_*\omega_{X/B}^{\otimes m} \neq 0$, then $f_*\omega_{X/B}^{\otimes m}$ is ample on B.*

Corollary 16.55. *Let $f : X \to B$ be an admissible family and $m > 1$. If $f_*\omega_{X/B}^{\otimes m} \neq 0$, then* $\deg f_*\omega_{X/B}^{\otimes m} > 0$.

The methods used to prove (16.54) give a more precise estimate of the positivity of these push-forwards as shown by Esnault and Viehweg:

Theorem 16.56 [EV90, 2.4]. *Let $f: X \to B$ be an admissible family, and $\mathscr{L}$ a line bundle on B. Assume that there exists an integer $m > 1$ such that $\deg \mathscr{L} < \deg f_*\omega_{X/B}^{\otimes m}$. Let r denote the rank of $f_*\omega_{X/B}^{\otimes m}$. Then there exists a positive integer $e = e(m,h)$, such that $(f_*\omega_{X/B}^{\otimes m})^{\otimes e\cdot r} \otimes \mathscr{L}^*$ is ample on B.*

Corollary 16.57 [Kov96, 2.15], [Kov00a, 2.1], [Kov02, 7.6]. (for $\Delta = \emptyset$) *Let $\mathscr{N}$ be a line bundle on B such that $\deg \mathscr{N}^{\otimes m\cdot e\cdot r} < \deg f_*\omega_{X/B}^{\otimes m}$ where $e = e(m,h)$ as above. Then $\omega_{X/B} \otimes f^*\mathscr{N}^*$ is ample on X.*

Proof (Sketch). As $\left(f_*(\omega_{X/B}^{\otimes m} \otimes f^*(\mathscr{N}^*)^{\otimes m})\right)^{\otimes e\cdot r} \simeq (f_*\omega_{X/B}^{\otimes m})^{\otimes e\cdot r} \otimes (\mathscr{N}^*)^{\otimes m\cdot e\cdot r}$, (16.56) implies that $f_*(\omega_{X/B}^{\otimes m} \otimes f^*(\mathscr{N}^*)^{\otimes m})$ is ample on B. Furthermore, by assumption one has that $\omega_{X/B}^{\otimes m} \otimes f^*(\mathscr{N}^*)^{\otimes m}|_{X_{\text{gen}}} \simeq \omega_{X_{\text{gen}}}^{\otimes m}$ is ample on X_{gen}. Hence $\omega_{X/B} \otimes f^*\mathscr{N}^*$ is ample both "horizontally" and "vertically", so it is ample. For details about the last step see [Kov02, 7.6]. □

This allows us to reduce the proof of (**WB**) to finding an appropriate line bundle on B according to the following plan.

PLAN 16.58. Find a line bundle $\mathscr{N}$ on B, depending only on B and Δ, such that $\omega_{X/B} \otimes f^*\mathscr{N}^*$ is not ample on X. Then we have $\deg \mathscr{N}^{\otimes m\cdot e\cdot r} \not< \deg f_*\omega_{X/B}^{\otimes m}$ by (16.57). In other words,

$$\deg f_*\omega_{X/B}^{\otimes m} \leq m \cdot e \cdot r \cdot \deg \mathscr{N}.$$

We find such an $\mathscr{N}$ using vanishing theorems. The main idea is the following: we want to find a line bundle such that twisting with the relative dualizing sheaf does not yield an ample line bundle. Ample line bundles appear in many vanishing theorems, so one way to prove that a given line bundle is not ample is to prove that a cohomology group does not vanish that would if the line bundle were ample. Next we are going to look at the needed vanishing theorems.

16.I.2 Vanishing theorems

We have already discussed some vanishing theorems in (3.H). As mentioned earlier, in order to prove (**WB**) we need a suitable vanishing theorem that is different from the ones we have already seen. The following is a somewhat weaker statement than what is really needed, but shows the main idea of the proof and how to apply it.

Theorem 16.59 [Kov97b, Kov00a]. *Let $f: X \to B$ be a family such that B is a smooth projective curve. Assume that $D = f^*\Delta$ is a normal crossing divisor. Let $n = \dim X_{\text{gen}}$ and $\mathscr{L}$ an ample line bundle on X such that $\mathscr{L} \otimes f^*(\omega_B(\Delta)^*)^{\otimes n}$ is also ample. Then*

$$H^{n+1}\left(X, \mathscr{L} \otimes f^*\omega_B(\Delta)\right) = 0.$$

Proof. After taking exterior powers of the sheaves of logarithmic differential forms, one has the following short exact sequence for each $p = 1, \dots, n+1$:

$$0 \longrightarrow \Omega^{p-1}_{X/B}(\log D) \otimes f^*\omega_B(\Delta) \longrightarrow \Omega^p_X(\log D) \longrightarrow \Omega^p_{X/B}(\log D) \longrightarrow 0.$$

Define $\mathscr{L}_p = \mathscr{L} \otimes f^*\omega_B(\Delta)^{\otimes 1-p}$ for $p = 0, \dots, n+1$. Then the above short exact sequence yields:

$$0 \longrightarrow \Omega^{p-1}_{X/B}(\log D) \otimes \mathscr{L}_{p-1} \longrightarrow \Omega^p_X(\log D) \otimes \mathscr{L}_p \longrightarrow \Omega^p_{X/B}(\log D) \otimes \mathscr{L}_p \longrightarrow 0.$$

$\mathscr{L}_p$ is ample for $p = 1, \dots, n+1$ since either $\omega_B(\Delta)$ or $\omega_B(\Delta)^*$ is nef. Then by (3.44) $H^{n+1-(p-1)}(X, \Omega^p_X(\log D) \otimes \mathscr{L}_p) = 0$ (recall that $\dim X = n+1$). Hence the map

$$H^{n+1-p}\left(X, \Omega^p_{X/B}(\log D) \otimes \mathscr{L}_p\right) \longrightarrow H^{n+1-(p-1)}\left(X, \Omega^{p-1}_{X/B}(\log D) \otimes \mathscr{L}_{p-1}\right)$$

is surjective for $p = 1, \dots, n+1$. Observe that these maps form a chain as p runs through $p = n+1, n, \dots, 1$. Hence the composite map,

$$H^0\left(X, \Omega^{n+1}_{X/B}(\log D) \otimes \mathscr{L}_{n+1}\right) \longrightarrow H^{n+1}(X, \mathscr{L}_0),$$

is also surjective. However, $\Omega^1_{X/B}(\log D)$ is of rank n, so $\Omega^{n+1}_{X/B}(\log D) = 0$, and therefore $H^{n+1}(X, \mathscr{L}_0) = H^{n+1}(X, \mathscr{L} \otimes f^*\omega_B(\Delta)) = 0$ as well. □

We are finally able to prove **(WB)**, at least for $\Delta = \emptyset$, by combining positivity and vanishing: (16.55) and (16.57) with $\mathscr{N} = \mathscr{O}_B$ imply that $\omega_{X/B}$ is ample. Since

$$H^{n+1}(X, \underbrace{\omega_{X/B} \otimes f^*\omega_B}_{\omega_X}) \neq 0,$$

this and (16.59) imply that $\omega_{X/B} \otimes (f^*\omega_B^*)^{\otimes n}$ cannot be ample. Then (16.57) with $\mathscr{N} = \omega_B^{\otimes n}$ implies that

$$\deg f_*\omega_{X/B}^{\otimes m} \leq \deg \omega_B^{\otimes n \cdot m \cdot e \cdot r} = m \cdot e \cdot r \cdot \dim X_{\text{gen}} \cdot (2g(B) - 2).$$

REMARK 16.60. For a complete proof of **(WB)** without the assumption $\Delta = \emptyset$, see [BV00], [Kov02], or [VZ02].

16.I.3 Kernels of Kodaira-Spencer maps

The essential part of the method described above was first used in [Kov96] and then it was polished through several articles [Kov97b, Kov97a, Kov00a, BV00, OV01, Kov02]. Finally, Viehweg and Zuo [VZ01, VZ02] combined some of the ideas of this method with Zuo's discovery of the negativity of kernels of Kodaira-Spencer maps [Zuo00]. This negativity is essentially a dual phenomenon of the positivity results mentioned earlier (16.54), (16.56).

The Viehweg-Zuo method has a great advantage over the previous method. The latter uses global vanishing theorems which limits the scope of the applications, while the Viehweg-Zuo method uses local arguments and hence is more applicable. Unfortunately this method is rather technical and so we cannot present it here. However, it is discussed in many places. The interested reader should start by consulting Viehweg's excellent survey [Vie01] and then reading the full account in [VZ01, VZ02]. Applications of these results of Viehweg and Zuo may be found in [KK07, KK08a, KK08b, KK08c].

16.J Allowing more general fibers

In the pursuit of more general results somewhat different approaches were taken by different authors. This includes setting up the generalized problem differently, using different techniques and not surprisingly obtaining somewhat different results. Here we discuss the approach of [Kov02] and other papers of Kovács. For a survey on the techniques used and results obtained in [VZ02] and subsequent papers of Viehweg and Zuo, the reader is referred to [Vie01] and the references therein.

Our starting point is a principle that has been applied with great success in birational geometry.

Principle 16.61. *Studying an ample line bundle on a singular variety is similar to studying a semi-ample and big line bundle on a smooth variety.*

The traditional way to use this principle is the following. The goal is to prove a statement for a pair, $(X, \mathscr{L})$, where X is possibly singular, and $\mathscr{L}$ is ample on X. Instead of working on X one works on a desingularization $f: Y \to X$, and considers the semi-ample and big line bundle $\mathscr{K} = f^*\mathscr{L}$. A prominent example of this trick is the use of the Kawamata-Viehweg vanishing theorem (3.40) in the minimal model program throughout Part II of this book.

Here we turn the situation upside-down. Our goal is a statement for $(Y, \mathscr{K})$, where Y is smooth and $\mathscr{K}$ is a semi-ample and big line bundle on Y. Instead of working on Y we construct a pair $(X, \mathscr{L})$ and a map $f: Y \to X$, where X is possibly singular, $\mathscr{L}$ is ample on X, f is birational, and $\mathscr{K} = f^*\mathscr{L}$.

The motivation for this approach is that we would like to extend the previous results to the case when $\omega_{X_{\text{gen}}}$ is not necessarily ample but only semi-ample and big. However, a crucial ingredient of the proof is an appropriate version of the Kodaira-Akizuki-Nakano vanishing theorem (3.42), and as Ramanujam (3.43) pointed out, (3.42) fails if the line bundle in question is only assumed to be semi-ample and big instead of ample. On the other hand, Navarro-Aznar *et al.* proved a singular version of the Kodaira-Akizuki-Nakano vanishing theorem, so one hopes that in this way a similar proof can work in this more general setting. For the statement of the theorem recall the definition of Du Bois's complex from (3.55).

Theorem 16.62 [Nav88, GNPP88]. *Let X be a complex projective variety and $\mathscr{L}$ an ample line bundle on X. Then*

$$\mathbb{H}^q(X, \underline{\Omega}^p_X \otimes \mathscr{L}) = 0 \qquad \text{for } p + q > \dim X.$$

Since Du Bois's complex agrees with the de Rham complex for smooth varieties, this theorem reduces to the Kodaira-Akizuki-Nakano theorem in the smooth case. However, this theorem is still not strong enough in our original situation if $\Delta \neq \emptyset$. We need a singular version of Esnault-Viehweg's logarithmic vanishing theorem (3.44).

Theorem 16.63 [Kov02, 4.1]. *Let X be a complex projective variety and $\mathscr{L}$ an ample line bundle on X. Further let D be an effective divisor on X. Then*

$$\mathbb{H}^q(X, \underline{\Omega}^p_X(\log D) \otimes \mathscr{L}) = 0 \qquad \text{for } p + q > \dim X.$$

To adapt the proof of (**WB**) to the singular case we need a singular version of (16.59). Besides the above vanishing theorem we also need an analog of the sheaf of relative logarithmic differentials.

16.64 Theorem-Definition [Kov02], cf. [Kov96, Kov97c, Kov05a]. *Let $f: X \to B$ be a morphism between complex varieties such that* $\dim X = n + 1$ *and B is a smooth curve. Let $\Delta \subseteq B$ be a finite set and $D = f^*\Delta$. For every non-negative integer p there exists a natural map $\wedge_p: \underline{\Omega}^p_X(\log D) \otimes f^*\omega_B(\Delta) \to \underline{\Omega}^{p+1}_X(\log D)$ and a complex $\underline{\Omega}^p_{X/B}(\log D) \in \mathrm{Ob}\,(D(X))$ with the following properties.*

(16.64.1) The natural map $\wedge_p$ factors through $\underline{\Omega}^p_{X/B}(\log D) \otimes f^*\omega_B(\Delta)$, i.e., there exist maps:

$$\begin{aligned} w''_p \ &: \underline{\Omega}^p_X(\log D) \otimes f^*\omega_B(\Delta) \to \underline{\Omega}^p_{X/B}(\log D) \otimes f^*\omega_B(\Delta) \quad \textit{and} \\ w'_p \ &: \underline{\Omega}^p_{X/B}(\log D) \otimes f^*\omega_B(\Delta) \to \underline{\Omega}^{p+1}_X(\log D) \end{aligned}$$

such that $\wedge_p = w'_p \circ w''_p$.

(16.64.2) If $w_p = w''_p \otimes id_{f^*\omega_B(\Delta)^*}: \underline{\Omega}^p_X(\log D) \to \underline{\Omega}^p_{X/B}(\log D)$, then

$$\underline{\Omega}^p_{X/B}(\log D) \otimes f^*\omega_B(\Delta) \xrightarrow{w'_p} \underline{\Omega}^{p+1}_X(\log D) \xrightarrow{w_{p+1}} \underline{\Omega}^{p+1}_{X/B}(\log D) \xrightarrow{+1}$$

is a distinguished triangle in $D(X)$.

(16.64.3) w_p is functorial, i.e., if $\phi: Y \to X$ is a B-morphism, then there are natural maps in $D(X)$ forming a commutative diagram:

$$\begin{array}{ccc} \underline{\Omega}^p_X(\log D) & \longrightarrow & \underline{\Omega}^p_{X/B}(\log D) \\ \downarrow & & \downarrow \\ R\phi_*\underline{\Omega}^p_Y(\log \phi^* D) & \longrightarrow & R\phi_*\underline{\Omega}^p_{Y/B}(\log \phi^* D). \end{array}$$

(16.64.4) $\underline{\Omega}^r_{X/B}(\log D) = 0$ for $r > n$.

(16.64.5) If f is proper, then $\underline{\Omega}^p_{X/B}(\log D) \in \mathrm{Ob}\left(D^b_{\mathrm{coh}}(X)\right)$ for every p.

(16.64.6) If f is smooth over $B \setminus \Delta$, then $\underline{\Omega}^p_{X/B}(\log D) \simeq_{\mathrm{qis}} \Omega^p_{X/B}(\log D)$.

Using these objects one can make the proof along the lines explained in (16.I.2) work to obtain the following theorem. It is in a non-explicit form. For more precise statements see [Kov02, (7.8), (7.10), (7.11), (7.13)].

Theorem 16.65. *Fix B and $\Delta \subset B$. Then weak boundedness holds for families of canonically polarized varieties with rational Gorenstein singularities and fixed Hilbert polynomial admitting a simultaneous resolution of singularities over $B \setminus \Delta$. In particular,*

$$2g(B) - 2 + \#\Delta > 0$$

for these families by (16.31).

As a corollary, one obtains weak boundedness for non-birationally-isotrivial families of minimal varieties of general type.

16.K Iterated Kodaira-Spencer maps and strong non-isotriviality

Let us finish this chapter by revisiting rigidity. We have seen in (16.F.1) that **(R)** fails as stated in the original conjecture and we asked

Question 16.66. Under what additional conditions does **(R)** hold?

This question was partially answered in [VZ03a] and [Kov05b]. Both papers gave essentially the same answer that we will discuss below. However, one must note that this is not the only case when rigidity holds as was shown in [VZ05a]. In other words we do not have a sufficient and necessary criterion for rigidity.

16.1 ITERATED KODAIRA-SPENCER MAPS, CASE I: ONE-DIMENSIONAL BASES. Let $f: X \to B$ be a smooth projective family of varieties of general type of dimension n, B a smooth (not necessarily projective) curve and let $T_X^m := \wedge^m T_X$ and $T_{X/B}^m := \wedge^m T_{X/B}$.

Let $1 \leq p \leq n$ and consider the short exact sequence,

$$0 \to T_{X/B}^p \otimes f^* T_B^{\otimes(n-p)} \to T_X^p \otimes f^* T_B^{\otimes(n-p)} \to T_{X/B}^{p-1} \otimes f^* T_B^{\otimes(n-p+1)} \to 0.$$

This induces an edge map,

$$\varrho_f^{(p)}: \mathcal{R}^{p-1} f_* T_{X/B}^{p-1} \otimes T_B^{\otimes(n-p+1)} \to \mathcal{R}^p f_* T_{X/B}^p \otimes T_B^{\otimes(n-p)}.$$

DEFINITION 16.67 [KOV05B]. Let $\varrho_f := \varrho_f^{(n)} \circ \varrho_f^{(n-1)} \circ \cdots \circ \varrho_f^{(1)}: T_B^{\otimes n} \longrightarrow \mathcal{R}^n f_* T_{X/B}^n$ and call f **strongly non-isotrivial** if $\varrho_f \neq 0$.

REMARK 16.68. The rationale behind this definition is that a family as above is isotrivial if and only if the Kodaira-Spencer map $\varrho_f^{(1)}$ is non-zero. The goal of *strong non-isotriviality* is to capture when a family is isotrivial in every direction on the *fiber*. To better understand this statement the reader is invited to analayze why rigidity fails for the example in (16.F.1).

EXAMPLE 16.69. Let $Y_i \to B$ be admissible families of curves for $i = 1, \ldots, r$. Then $X = Y_1 \times_B \cdots \times_B Y_r \to B$ is strongly non-isotrivial.

REMARK 16.70. Since T_B is a line bundle and $\mathscr{R}^n f_* T^n_{X/B}$ is locally free, $\varrho_f \neq 0$ if and only if it is injective. We use this observation in the definition of strong non-isotriviality for higher dimensional bases.

16.2 ITERATED KODAIRA-SPENCER MAPS, CASE II: HIGHER-DIMENSIONAL BASES. Let $f : X \to B$ be a smooth projective family of varieties of general type of dimension n, B a smooth (not necessarily projective) variety.

For an integer p, $1 \leq p \leq n$, there exists a filtration

$$T^p_X = \mathscr{F}^0 \supseteq \mathscr{F}^1 \supseteq \cdots \supseteq \mathscr{F}^p \supseteq \mathscr{F}^{p+1} = 0,$$

such that

$$\mathscr{F}^i \Big/ \mathscr{F}^{i+1} \simeq T^i_{X/B} \otimes f^* T^{p-i}_B.$$

In particular,

$$\mathscr{F}^p \simeq T^p_{X/B}$$

and

$$\mathscr{F}^{p-1} / \mathscr{F}^p \simeq T^{p-1}_{X/B} \otimes f^* T_B.$$

Therefore one has a short exact sequence,

$$0 \to T^p_{X/B} \otimes f^* T_B^{\otimes(n-p)} \to \mathscr{F}^{p-1} \otimes f^* T_B^{\otimes(n-p)} \to T^{p-1}_{X/B} \otimes f^* T_B^{\otimes(n-p+1)} \to 0,$$

that induces a map

$$\varrho^{(p)}_f : \mathscr{R}^{p-1} f_* T^{p-1}_{X/B} \otimes T_B^{\otimes(n-p+1)} \to \mathscr{R}^p f_* T^p_{X/B} \otimes T_B^{\otimes(n-p)}.$$

DEFINITION 16.71 [KOV05B]. Let $\varrho_f : \operatorname{Sym}^n T_B \longrightarrow \mathscr{R}^n f_* T^n_{X/B}$ be the natural map induced by $\varrho^{(n)}_f \circ \varrho^{(n-1)}_f \circ \cdots \circ \varrho^{(1)}_f$ and call f **strongly non-isotrivial over** B if ϱ_f is injective.

EXAMPLE 16.72. Let $Y_i \to B$ be non-isotrivial families of smooth projective curves for $i = 1, \ldots, r$. Then $X = Y_1 \times_B \cdots \times_B Y_r \to B$ is strongly non-isotrivial over B.

REMARK 16.73. One could consider various refinements of this notion. For instance, consider maps for which the composition of fewer $\varrho^{(p)}$'s is injective or non-zero. These appear for example in the study of moduli spaces of varieties that are products with one rigid term. One could also combine this condition with $\operatorname{Var} f$, the variation of f in moduli. This is a mostly unexplored area at the moment.

A partial possible answer to (16.35) is given by the following theorem:

Theorem 16.74 [VZ03a], [Kov05b]. *Let $f : X \to B$ be a smooth projective family of varieties of general type, B a smooth variety. If f is strongly non-isotrivial over B, then rigidity holds for f.*

This, combined with (16.40), leads to a statement resembling the original Shafarevich conjecture. In fact, for families of curves it reduces to the original statement.

Theorem 16.75 [KL06]. *Let B, Δ and h be fixed. Then there exist only finitely many strongly non-isotrivial families of canonically polarized varieties with Hilbert polynomial h with respect to B and Δ.*

Part IV

Solutions and hints to some of the exercises

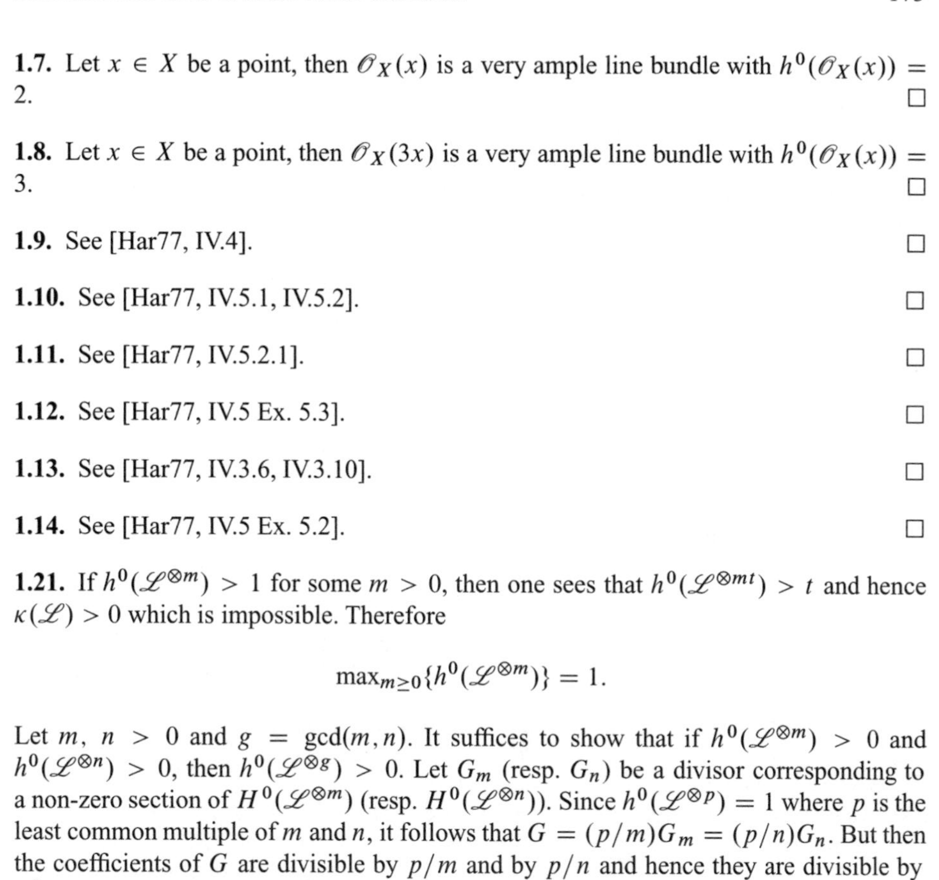

1.7. Let $x \in X$ be a point, then $\mathscr{O}_X(x)$ is a very ample line bundle with $h^0(\mathscr{O}_X(x)) = 2$. □

1.8. Let $x \in X$ be a point, then $\mathscr{O}_X(3x)$ is a very ample line bundle with $h^0(\mathscr{O}_X(x)) = 3$. □

1.9. See [Har77, IV.4]. □

1.10. See [Har77, IV.5.1, IV.5.2]. □

1.11. See [Har77, IV.5.2.1]. □

1.12. See [Har77, IV.5 Ex. 5.3]. □

1.13. See [Har77, IV.3.6, IV.3.10]. □

1.14. See [Har77, IV.5 Ex. 5.2]. □

1.21. If $h^0(\mathscr{L}^{\otimes m}) > 1$ for some $m > 0$, then one sees that $h^0(\mathscr{L}^{\otimes mt}) > t$ and hence $\kappa(\mathscr{L}) > 0$ which is impossible. Therefore

$$\max_{m \geq 0}\{h^0(\mathscr{L}^{\otimes m})\} = 1.$$

Let $m,\ n > 0$ and $g = \gcd(m,n)$. It suffices to show that if $h^0(\mathscr{L}^{\otimes m}) > 0$ and $h^0(\mathscr{L}^{\otimes n}) > 0$, then $h^0(\mathscr{L}^{\otimes g}) > 0$. Let G_m (resp. G_n) be a divisor corresponding to a non-zero section of $H^0(\mathscr{L}^{\otimes m})$ (resp. $H^0(\mathscr{L}^{\otimes n})$). Since $h^0(\mathscr{L}^{\otimes p}) = 1$ where p is the least common multiple of m and n, it follows that $G = (p/m)G_m = (p/n)G_n$. But then the coefficients of G are divisible by p/m and by p/n and hence they are divisible by p^2/mn. But then, since $p \cdot \frac{mn}{p^2} = \frac{mn}{p} = g$, we have a divisor $\frac{mn}{p^2}G$ corresponding to a non-zero section of $H^0(\mathscr{L}^{\otimes g})$. □

1.22. See [Laz04a, 2.1.38]. □

2.1. If $A \sim_{\mathbb{R}} B$, then we have $A - B = \sum r_i(f_i)$ where $r_i \in \mathbb{R}$ and f_i are rational functions on X. Since the r_i are a real-valued solution to a system of equations with rational coefficients, then there exists a rational solution to the same system of equations, i.e., there are numbers $q_i \in \mathbb{Q}$ such that $A - B = \sum q_i(f_i)$. Therefore $A \sim_{\mathbb{Q}} B$. □

2.2. By Noetherian induction, we have $\cap_{n>0}\mathrm{Bs}(nD) = \mathrm{Bs}(mD)$ for any $m > 0$ sufficiently divisible. The inclusion $\mathbf{B}(D) \subset \mathrm{Bs}(mD)$ is clear. If $x \notin \mathbf{B}(D)$, then there is an $\mathbb{R}$-divisor $D' \geq 0$ such that $D \sim_{\mathbb{R}} D'$ and $x \notin \mathrm{Supp}(D')$. We have $D' - D = \sum r_i(f_i)$ where $r_i \in \mathbb{R}$ and f_i are rational functions on X. We may assume that $\mathrm{Supp}(f_i) \subset \mathrm{Supp}(D' - D)$ (e.g., by assuming that r_i are linearly independent over $\mathbb{Q}$ cf. (2.13)). Then if q_i are rational numbers with $|q_i - r_i| \ll 1$, we have that $D'' = D + \sum q_i(f_i)$ is a $\mathbb{Q}$-divisor $\mathbb{Q}$-linearly equivalent to D with the same support as D'. If $m > 0$ is sufficiently divisible, it then follows that $x \notin \mathrm{Bs}(mD)$ as required. □

2.4. Replacing D by a multiple (cf. (5.68)), we may assume that $|D|$ is basepoint-free and so the map $f : X \to \mathbb{P}^n = \mathbb{P}H^0(\mathscr{O}_X(D))$ is a morphism and $\mathscr{O}_X(D) = f^*\mathscr{O}_{\mathbb{P}^N}(1)$. We may also assume that f is surjective with connected fibers onto a subvariety $Z \subset \mathbb{P}^N$. Therefore, by the projection formula, $H^0(\mathscr{O}_X(kD)) \simeq H^0(\mathscr{O}_Z(k))$ for all $k > 0$. By Serre vanishing, there exists an integer $r > 0$ such that $H^0(\mathscr{O}_{\mathbb{P}^N}(rk)) \to H^0(\mathscr{O}_Z(rk))$ is onto for all $k \geq 0$. Since $\oplus_{k\geq 0} H^0(\mathscr{O}_{\mathbb{P}^N}(rk))$ is finitely generated,

$$\oplus_{k\geq 0} H^0(\mathscr{O}_X(rkD)) \simeq \oplus_{k\geq 0} H^0(\mathscr{O}_Z(rk))$$

is finitely generated. By (5.68) $R(D)$ is finitely generated. □

2.5. See [KM98, 2.53]. □

2.6. See [KM98, 2.49]. □

2.7. Let $\vartheta : \mathscr{O}_X \to \mathscr{L}^{[m]}$ be a section and make the same construction with this choice of ϑ as in (2.D). Outside the zero locus of ϑ the resulting cover is the same as before. See [KM98, 2.52] for more details. □

2.11. If D is big, then $\limsup \left(h^0(\mathscr{O}_X(mD))/m^{\dim X}\right) > 0$. Since $h^0(\mathscr{O}_Y(mf^*D)) \geq h^0(\mathscr{O}_X(mD))$, it follows that f^*D is big.

If F is an effective divisor whose support contains the exceptional divisors of f, then there exists a $k > 0$ such that $f^*D \leq f_*^{-1}D + kF$. It follows that

$$h^0(\mathscr{O}_X(mD)) \leq h^0(\mathscr{O}_Y(mf^*D)) \leq h^0(\mathscr{O}_Y(mkf_*^{-1}D + F)).$$

Therefore, if D is big, $f_*^{-1}D + F$ is also big.

If $G \sim G'$, then there are integers n_i and rational functions f_i on Y (and hence on X) such that $G - G' = \sum n_i(f_i)$. Let $X^0 = X - f(\mathrm{Ex}(f))$, then $(G - G')|_{X^0} = \sum n_i(f_i)|_{X^0}$. Since $X - X^0$ has codimension ≥ 2 in X, we have that $f_*G - f_*G' = \sum n_i(f_i)$ and so $f_*G \sim f_*G'$. If G is big, then

$$\limsup \left(h^0(\mathscr{O}_X(mf_*G))/m^{\dim X}\right) \geq \limsup \left(h^0(\mathscr{O}_Y(mG))/m^{\dim Y}\right) > 0.$$

Let X be the surface obtained by blowing up $\mathbb{P}^2$ at a point and G be the strict transform of a line $f_*G \subset \mathbb{P}^2$. Then G is a fiber of the corresponding morphism $X \to \mathbb{P}^1$ and so G is not big, but f_*G is big. □

2.12. See [BCHM10, §3]. □

2.13. See [BCHM10, §3]. □

2.14. If $|G|_{\mathbb{Q}} \neq 0$, then there are rational numbers q_i and rational functions f_i such that $G + \sum q_i(f_i) \geq 0$. Therefore, there is an integer $m > 0$ such that $mq_i \in \mathbb{Z}$ and $mG + \sum mq_i(f_i) \geq 0$, i.e., $|mG| \neq \emptyset$.

Let $G = \sqrt{2}(p-q)$ where p and q are distinct points on $\mathbb{P}^1$. Then $H^0(\mathscr{O}_{\mathbb{P}^1}(mG)) = H^0(\mathscr{O}_{\mathbb{P}^1}(\lfloor mG \rfloor))$. Since

$$\deg(\lfloor mG \rfloor) = \left\lfloor m\sqrt{2} \right\rfloor + \left\lfloor -m\sqrt{2} \right\rfloor < 0 \qquad \forall m > 0,$$

we have $H^0(\mathcal{O}_{\mathbb{P}^1}(mG)) = 0$ for all $m > 0$. □

3.3. By the adjunction formula we have that

$$K_X = (K_W + X)\big|_X,$$
$$K_{X_1} = (K_W + X_1)\big|_{X_1},$$
$$K_{X_2} = (K_W + X_2)\big|_{X_2}.$$

Therefore

$$K_X\big|_{X_i} = K_{X_i} + X_{3-i}\big|_{X_i}.$$ □

3.4. By the assumption on the Kodaira dimension there exists an $m > 0$ such that mK_Z is effective, hence so is $m(K_Z + \Gamma)$. Then by the assumption on the intersection numbers, $(K_Z + \Gamma)^2 > 0$, so the statement follows by the Nakai-Moishezon criterium. □

3.5. Let $f : Y \to X$ be the resolution given by blowing up the vertex $P \in C_X$ and $E \simeq X$ be the exceptional divisor. If C_D is $\mathbb{Q}$-Cartier, then $f^*C_D \sim_{\mathbb{Q}} f_*^{-1}C_D + \lambda E$ so that $0 \sim_{\mathbb{Q}} f^*C_D|_E \sim D + \lambda c_1 \mathcal{O}_X(-1)$. □

3.12. It follows directly from the definition of canonical singularities and the fact that ω_X is a line bundle that $f^*\omega_X \hookrightarrow \omega_{\tilde{X}}$. This implies (3.12.1). Pushing this morphism forward we obtain a morphism $\tau : \omega_X \to f_*f^*\omega_X \to f_*\omega_{\tilde{X}}$ which is an isomorphism on the complement of a codimension 2 subvariety. Since $f_*\omega_{\tilde{X}}$ is a torsion free sheaf, we obtain that τ is an injection and since ω_X is reflexive and τ is an isomorphism on the complement of a codimension 2 subvariety, it follows that τ is an isomorphism. This implies (3.12.2). □

3.17. See [Har77, Ex.III.3.5]. □

3.27. Let $f : Y \to X$ be the blow-up of the vertex $O \in X$ and $E \subset Y$ the corresponding exceptional divisor. Then $E \simeq \mathbb{P}^1$ and $E^2 = -n$. We write $K_Y = f^*K_X + aE$. By adjunction

$$-2 = \deg(K_{\mathbb{P}^1}) = \deg(K_Y + E)|_E = (K_Y + E) \cdot E = (a+1)E^2 = -n(a+1).$$

Therefore $a = -1 + 2/n$. □

3.28. Let $f : Y \to X$ be the blow-up of the vertex $O \in X$ and $E \subset Y$ the corresponding exceptional divisor. Then E is an elliptic curve and $E^2 = -n$ for some $n > 0$. By adjunction

$$0 = \deg(K_E) = \deg(K_Y + E)|_E = (K_Y + E) \cdot E = (a+1)E^2 = -n(a+1).$$

Therefore $a = -1$ and X is lc but not klt. Notice that $\mathcal{R}^1 f_*\mathcal{O}_Y \simeq \mathcal{R}^1 f_*\mathcal{O}_E = \mathbb{C}_O$ and so X does not have rational singularities. □

3.29. Let $f : Y \to X$ be the blow-up of the vertex $O \in X$ and $E \subset Y$ the corresponding exceptional divisor. Then $E \simeq \mathbb{P}^1$ and $E^2 = -n$ and $K_Y = f^*K_X + (-1 + 2/n)E$ cf. (3.27). Let $T = f_*^{-1}S$ and write $f^*S = T + bE$. We have that $0 = (T + bE) \cdot E = 1 - nb$ so that $b = 1/n$. Therefore $f^*(K_X + S) = K_Y + T + (1 - 2/n + 1/n)E$. As $1 - 1/n = 1 - 2/n + 1/n < 1$, it follows that (X, S) is plt and $(K_X + S)|_S \equiv (K_Y + T)|_T \equiv K_T + (1 - 1/n)O$. □

3.30. If we blow up the vertex O of the cusp, then strict transform of the cusp is smooth and intersects the exceptional divisor in a point P. If we blow up P, then the strict transform of the cusp, the strict transform of the first exceptional divisor and the new exceptional divisor meet at a point Q (and each pair of divisors in fact meets transversely). Let $f : Y \to X$ be the morphism obtained by blowing up O, P and Q, then f is a log resolution of (X, S). If S', E_1, E_2 and E_3 denote the strict transforms of S, of the first and second exceptional divisors and the third exceptional divisor, then $f^*S = S' + 2E_1 + 3E_2 + 6E_3$ and $K_{Y/X} = E_1 + 2E_2 + 4E_3$ so that $f^*(K_X + tS) = K_Y + tS' + (2t - 1)E_1 + (3t - 2)E_2 + (6t - 4)E_3$. If $t < 5/6$ then each coefficient is < 1 and if $t \geq 5/6$, then $6t - 4 \geq 1$. □

3.31. Blow up smooth divisors contained in F and its strict transforms. □

3.32. Assume discrep.$(X, \Delta) \neq -\infty$, then discrep.$(X, \Delta) \geq -1$ by (3.31). The inequality totaldiscrep.$(X, \Delta) \leq$ discrep.(X, Δ) is clear. Let $P \in X$ be a smooth codimension 2 point not contained in Supp(Δ) and E be the exceptional divisor given by blowing up $P \in X$. Then $a(E; X, \Delta) = 1$ and so the last inequality follows. □

3.33. See [KM98, 2.31]. □

3.34. See [KM98, 2.43]. □

3.51. Let $A = L - (K_Y + \Gamma)$. There exists an effective divisor $\Gamma' \sim_{\mathbb{R},X} \Gamma + A/2$ such that (Y, Γ') is klt and $L - (K_Y + \Gamma') \sim_{\mathbb{R},X} A/2$ is f-ample. □

5.7. See [KM98, §3]. □

5.8. Since Δ is big and (X, Δ) is klt, there exists an $\mathbb{R}$-divisor $\Delta' = A + B \sim_{\mathbb{R}} \Delta$ such that (X, Δ') is klt, $B' \geq 0$ and A is ample. The claim now follows from 2 of (5.4). □

5.9. We may write $aD - (K_X + \Delta) \sim_{\mathbb{R},Z} A + B$ where $B \geq 0$ is an $\mathbb{R}$-divisor and A is an ample $\mathbb{Q}$-divisor. Note that then B is $\mathbb{R}$-Cartier and so for any rational number $0 < \varepsilon \ll 1$ the pair $(X, \Delta + \varepsilon(B + A/2))$ is klt. It follows that

$$aD - (K_X + \Delta + \varepsilon(B + A/2)) \sim_{\mathbb{R},Z} (1 - \varepsilon)(A + B) + \varepsilon A/2$$

is ample over Z. Let $\Delta' \in V$ be any $\mathbb{Q}$-divisor sufficiently close to $\Delta + \varepsilon(B + A/2)$ such that $K_X + \Delta'$ is $\mathbb{Q}$-Cartier. Then the pair (X, Δ') is klt and $aD - (K_X + \Delta')$ is ample over Z and Δ' is big over Z. By (5.8) there are finitely many negative extremal rays for $K_X + \Delta'$ over Z. Let $V \subset \mathrm{WDiv}_{\mathbb{R}}(X)$ be the smallest rational affine subspace containing

aD. If $aD' \in V$ and $||D - D'|| \ll 1$, then $aD' - (K_X + \Delta')$ is ample over Z. For any curve $C \subset X$, we have

$$(K_X + \Delta') \cdot C = -(aD' - K_X - \Delta') \cdot C + aD' \cdot C < aD' \cdot C.$$

Therefore, if aD' is not nef over Z, then there is a $(K_X + \Delta')$-negative extremal ray R_i over Z. Notice that as $aD' \in V$, if $aD \cdot R_i = 0$, then $aD' \cdot R_i = 0$ and as $||D - D'|| \ll 1$, if $aD \cdot R_i > 0$, then $aD' \cdot R_i > 0$. It follows that aD' is nef over Z. We may then write $aD = \sum r_i D_i$ where $r_i \in \mathbb{R}_{>0}$, $D_i \in V \cap \mathrm{WDiv}_{\mathbb{Q}}(X)$ and $||aD - D_i|| \ll 1$. Therefore $D_i - (K_X + \Delta')$ is ample over Z and D_i is nef over Z. But then it follows by (5.1) that D_i is semiample over Z. Let $f_i : X \to Y_i$ be the corresponding morphism over Z, so that $D_i \sim_{\mathbb{Q},Z} f_i^* H_i$ where H_i is an ample $\mathbb{Q}$-divisor over Z. Then f_i is determined by the set of D_i trivial curves over Z, i.e., by the face $F_i = \overline{NE}(X/Z)_{(D_i)=0}$. As observed above, F_i is spanned by finitely many $K_X + \Delta'$ negative extremal rays over Z and F_i does not depend on i. Therefore $aD \sim_{\mathbb{R},Z} \sum r_i f^* H_i$ where $f = f_i : X \to Y = Y_i$, i.e., D is semiample over Z. □

5.10. We may write $A \sim_{\mathbb{R}} A' + C$ where A' is a general ample $\mathbb{Q}$-divisor and C is an effective $\mathbb{R}$-divisor whose support contains no non-klt center of (X, Δ). For $0 < \varepsilon \ll 1$, we then have that $(X, \Delta' = (1 - \varepsilon)\Delta + \varepsilon C)$ is klt cf. (3.34), $D = K_X + \Delta$ is nef and $D - (K_X + \Delta')$ is ample. The claim now follows from (5.1). □

5.11. This follows easily from (5.9). □

5.12. See [Laz09, 1.5.4]. □

5.13. See [CKM88, 5.1]. □

5.31. By (5.24) and (5.25), the X_i are $\mathbb{Q}$-factorial and $a(E; X_i, \Delta_i) \le a(E; X_{i+1}, \Delta_{i+1})$ where the strict inequality holds if E is $X_i \dashrightarrow X_{i+1}$ exceptional. □

5.33. If $K_Z + f_*\Delta$ is $\mathbb{Q}$-Cartier, then as f is small, $K_X + \Delta = f^*(K_Z + f_*\Delta)$. This contradicts the fact that $-(K_X + \Delta)$ is ample over Z. □

5.34. Let E_1 and E_2 be distinct divisorial components of $\mathrm{Ex}(f)$. Since $\varrho(X/Z) = 1$, we have $E_1 + aE_2 \equiv_Z 0$ for some $0 \neq a \in \mathbb{R}$. Each E_i is covered by f exceptional curves C_i such that $C_i \cdot E_i < 0$. If we may pick C_1 not intersecting E_2, then $0 = (E_1 + aE_2) \cdot C_1 = E_1 \cdot C_1 < 0$ which is impossible. Therefore, we may assume that C_1 intersects E_2 and $a = -(E_1 \cdot C_1)/(E_2 \cdot C_2) > 0$. But since $E = E_1 + aE_2 \ge 0$, then E is covered by curves C over Z with $E \cdot C < 0$. This contradicts the fact that $E \equiv_Z 0$. □

5.35. Let X^0 and Y^0 be open sets such that $f|_{X^0} : X^0 \dashrightarrow Y^0$ is an isomorphism and the codimension of $X - X^0$ in X and $Y - Y^0$ in Y is at least 2. Then

$$\begin{aligned} H^0(\mathcal{O}_X(D)) &= \{g \in K(X) | D + (g) \ge 0\} = \{g \in K(X^0) | \, D|_{X^0} + (g) \ge 0\} \\ &= \{g \in K(Y^0) | \, D|_{Y^0} + (g) \ge 0\} = \{g \in K(Y) | f_*D + (g) \ge 0\} = H^0(\mathcal{O}_Y(f_*D)). \end{aligned}$$

□

5.36. Let $\phi\colon X \dashrightarrow Y$ be an ltm for (X, Δ) over Z. Then $(Y, \phi_*\Delta)$ is klt and $\phi_*\Delta$ is big. By the Basepoint-free theorem (5.1), it follows that $K_Y + \phi_*\Delta$ is semiample over Z. Let $p : W \to X$ and $q : W \to Y$ be a common resolution. Then

$$p^*(K_X + \Delta) = q^*(K_Y + \phi_*\Delta) + F$$

where $F \geq 0$ is a q-exceptional $\mathbb{Q}$-divisor whose support contains the divisors contracted by ϕ. Since $\mathbf{B}(p^*(K_X + \Delta)/Z) = \mathbf{B}(q^*(K_Y + \phi_*\Delta) + F/Z) = \mathrm{Supp}(F)$, the assertion now follows easily. □

5.37. Let $p : W \to X$ and $q_i : W \to Y_i$ be a common resolution. Then $p^*(K_X + \Delta) = q_i^*(K_{Y_i} + \phi_{i*}\Delta) + E_i$ where $E_i \geq 0$ is q_i-exceptional and p_*E_i is supported on the set of ϕ_i-exceptional divisors. Since $q_1^*(K_{Y_1} + \phi_{1*}\Delta) = q_2^*(K_{Y_2} + \phi_{2*}\Delta) + E_2 - E_1$, $q_2^*(K_{Y_2} + \phi_{2*}\Delta)$ is nef over Y_1 and $q_{1*}(E_2 - E_1) \geq 0$, by (5.23), $E_2 - E_1 \geq 0$. By symmetry $E_1 = E_2$ and the statement follows. □

5.38. The singularities of W are locally isomorphic to cones over the Veronese surface S given by embedding $\mathbb{P}^2$ in $\mathbb{P}^5$ via the complete linear series corresponding to $H^0(\mathcal{O}_{\mathbb{P}^2}(2))$. These can be resolved by blowing up the vertices $b : W' \to W$. We then have that the exceptional divisor $E_i \subset W'$ corresponding to the singular point $w_i \in W$ satisfies $\mathcal{O}_{E_i}(E_i) \simeq \mathcal{O}_{\mathbb{P}^2}(-2)$. By adjunction, $\mathcal{O}_{E_i}(K_{W'} + E_i) \simeq \mathcal{O}_{\mathbb{P}^2}(-3)$. The morphism b is given by a sequence of divisorial contractions and hence K_W is $\mathbb{Q}$-Cartier (this can also be checked directly). Let l_i be a line in E_i. We then have $0 = b^*K_W \cdot l_i$ and so $K_{W'} = b^*K_W + \frac{1}{2}\sum E_i$. Since $(W', \sum E_i)$ has simple normal crossings, b is a log resolution and W has terminal singularities. If $(W^m, 0)$ is any ltm of $(W, 0)$, then by (5.37), W^m is isomorphic to W in codimension 1. If W^m were smooth, then $a(E_i; W^m, 0)$ would be an integer. This is impossible. □

5.39. Let $\phi_i\colon X_i \dashrightarrow X_{i+1}$ be a sequence of flips and divisorial contractions for $K_X + \Delta$. Since $K_X + \Delta \equiv \lambda(K_X + \Delta')$ for some $\lambda > 0$, one sees that $K_{X_i} + \Delta_i \equiv \lambda(K_{X_i} + \Delta_i')$. But then $X_i \to Z_i$ is a $K_{X_i} + \Delta_i$ flip (resp. divisorial contraction) if and only if it is a $K_{X_i} + \Delta_i'$ flip (resp. divisorial contraction). □

5.40. Similar to (5.39). □

5.41. See [Kol92, §17]. □

5.42. Let $\phi : Y \dashrightarrow Z$ be an ltm over U for $K_Y + \Gamma + F$. One sees that $\mathbf{B}_-(K_Y + \Gamma + F/U)$ contains the support of F which is therefore contracted by ϕ. It follows that $\psi := \phi \circ f^{-1} : X \dashrightarrow Z$ extracts no divisors and $\psi_*(K_X + \Delta) = \phi_*(K_Y + \Gamma + F)$ is nef over U so that ψ is an ltm for $K_X + \Delta$ over U. □

5.43. Let C be a general ample divisor on X'. We may assume that $\mathbf{B}(A - f_*^{-1}C)$ contains no non-klt centers of (X, Δ) so that $A - f_*^{-1}C \sim_{\mathbb{Q}} D \geq 0$ where $(X, \Delta' = \Delta + \varepsilon D)$ is dlt for $0 < \varepsilon \ll 1$. But then f is also a $(K_X + \Delta')$-flip and so $(X', f_*\Delta' = f_*\Delta + \varepsilon f_*D)$ is dlt cf. (5.24). Therefore the support of f_*D contains no non-klt centers of $(X', f_*\Delta)$. Since $A' = f_*A \sim_{\mathbb{Q}} f_*D + C$, $\mathbf{B}_+(A')$ contains no non-klt centers of (X', Δ'). □

5.44. Let $\phi : X \dashrightarrow X^+$ be the flip of f. If C^+ is a flipped curve, then there is an exceptional divisor E with center C^+ such that $a(E;X) < a(E;X^+) = 1$. If X is smooth, then $a(E;X)$ is an integer and hence $a(E;X) \leq 0$. This is a contradiction (as X is terminal). $\square$

5.45. Y is smooth over X_{reg} and the exceptional divisor E maps to a subvariety $V \subset X_{\mathrm{reg}}$ of codimension $b+1 \geq 2$. Let $v \in V$ and $E_v = E \times_v V$. Then $E_v \simeq \mathbb{P}^b$ and a line l on E_v generates an extremal ray of $\overline{\mathrm{NE}}(Y)$ (in fact, if H is ample on X, then $l \cdot f^*H = 0$ and $C \cdot f^*H > 0$ for any curve such that $[C] \notin \mathbb{R}_{\geq 0}[l]$). Since $\mathcal{O}_{E_v}(E) = \mathcal{O}_{\mathbb{P}^b}(-1)$ and $\mathcal{O}_{E_v}(K_{V/Y} + E) = \mathcal{O}_{\mathbb{P}^b}(-b-1)$, we have $K_{V/Y} \cdot l = -b < 0$. $\square$

5.46. $p : P \to X$ is a surjective morphism with connected fibers such that $\varrho(P/X) = 1$. Let F be a fiber, then it is easy to see that F generates an extremal ray of $\overline{\mathrm{NE}}(P)$ and $K_Y \cdot F = -2 < 0$. Therefore, $P \to X$ is a Mori fiber space. If $\mathcal{L} = \mathcal{O}_{\mathbb{P}^n}(a)$, then $\mathcal{O}_{X_0}(X_0) \simeq \mathcal{O}_{\mathbb{P}^n}(-a)$ and $K_P = \mathcal{O}_P(-2X_0) \otimes p^*\mathcal{O}_{\mathbb{P}^n}(-a-n-1)$. Let l be a line on X_0, then $K_P \cdot l = a - n - 1 < 0$. If C is a curve not contained in X_0, then $C \cdot (X_0 + p^*\mathcal{O}_{\mathbb{P}^n}(a)) > 0$ and if $C \subset X_0$ then $C \cdot (X_0 + p^*\mathcal{O}_{\mathbb{P}^n}(a)) = 0$. Therefore $[l]$ spans a K_P-negative extremal ray and hence it defines a divisorial contraction. $\square$

5.47. $f : W \to V$ is given by blowing up the vertex. The exceptional divisor E is then isomorphic to the quadric defined by $\{wz - xy = 0\} \subset \mathbb{P}^3$. We have $\mathcal{O}_E(E) \simeq \mathcal{O}_{\mathbb{P}^1\times\mathbb{P}^1}(-1,-1)$. There is a projection $p : W \to \mathbb{P}^1 \times \mathbb{P}^1$ and W is a $\mathbb{P}^1$ bundle over $\mathbb{P}^1 \times \mathbb{P}^1$ defined by the vector bundle $\mathcal{O}_{\mathbb{P}^1\times\mathbb{P}^1} \oplus \mathcal{O}_{\mathbb{P}^1\times\mathbb{P}^1}(-1,-1)$. We have that E is the negative section corresponding to the unique surjection $\mathcal{O}_{\mathbb{P}^1\times\mathbb{P}^1} \oplus \mathcal{O}_{\mathbb{P}^1\times\mathbb{P}^1}(-1,-1) \to \mathcal{O}_{\mathbb{P}^1\times\mathbb{P}^1}(-1,-1)$. Let F be the section corresponding to a general surjection $\mathcal{O}_{\mathbb{P}^1\times\mathbb{P}^1} \oplus \mathcal{O}_{\mathbb{P}^1\times\mathbb{P}^1}(-1,-1) \to \mathcal{O}_{\mathbb{P}^1\times\mathbb{P}^1}$. Then $|F|$ is free and, if $\mathcal{L}_1 = p^*\mathcal{O}_{\mathbb{P}^1\times\mathbb{P}^1}(1,2) \otimes \mathcal{O}_W(E)$, then $\mathcal{L}_1$ is generated. Indeed, $\mathcal{L}_1 = \mathcal{O}_{\mathbb{P}^1\times\mathbb{P}^1}(0,1) \otimes \mathcal{O}_W(F)$ is the the tensor product of two generated line bundles. If $C \subset W$ is a curve, then we have that $\mathcal{L}_1 \cdot C = 0$ iff $C \subset E$ is a ruling for the first projection $E \to \mathbb{P}^1$ and so $f_1 : W \to W_1 \subset \mathbb{P}H^0(\mathcal{L}_1)$ is a divisorial contraction over V mapping E to a curve $D_1 \subset W_1$. Let $\mathrm{Cl}(V)$ be the group of divisors on V modulo linear equivalence. Then $\mathrm{Cl}(V)$ is generated by divisors C_D given by cones over the corresponding divisor $D \in \mathbb{P}^1 \times \mathbb{P}^1$. A divisor C_D is $\mathbb{Q}$-Cartier iff the class of D is a multiple of the class of $\mathcal{O}_{\mathbb{P}^1\times\mathbb{P}^1}(1,1)$ and so as $K_V = C_{K_{\mathbb{P}^1\times\mathbb{P}^1}}$, K_V is $\mathbb{Q}$-Cartier. Since $W_1 \to V$ is small $K_{W_1} = g_1^*K_V$ where $g_1 : W_1 \to V$ is the induced morphism and so g_1 is K_{W_1}-trivial. A similar computation holds for $\mathcal{L}_2 = p^*\mathcal{O}_{\mathbb{P}^1\times\mathbb{P}^1}(2,1) \otimes \mathcal{O}_W(E)$ yielding $W \to W_2 \to V$. We let Δ_1 be the divisor associated to a small multiple of $f_{1*}\mathcal{L}_2$. $\square$

5.48. This is similar to (5.47), however K_Z is not $\mathbb{Q}$-Cartier. We let $f_i : W \to W_i \subset \mathbb{P}H^0(\mathcal{L}_i^{\otimes s})$ where $p : W \to \mathbb{P}^1 \times \mathbb{P}^1$ is the projection and $\mathcal{L}_1 = p^*\mathcal{O}_{\mathbb{P}^1\times\mathbb{P}^1}(2,3) \otimes \mathcal{O}_W(E)$ and $\mathcal{L}_2 = p^*\mathcal{O}_{\mathbb{P}^1\times\mathbb{P}^1}(1,4) \otimes \mathcal{O}_W(E)$. Since $\mathcal{O}_E(E) = \mathcal{O}_{\mathbb{P}^1\times\mathbb{P}^1}(-1,-3)$, the morphism f_i contracts the $(3-i)$-th ruling of E. We have that $\mathcal{O}_E(K_W) = \mathcal{O}_E(K_E - E) = \mathcal{O}_{\mathbb{P}^1\times\mathbb{P}^1}(-1,1)$ so that $K_W + E/3 = f_1^*K_{W_1}$ and $K_W - E = f_2^*K_{W_2}$. It follows that $a(F;W_1) < a(F;W_2)$ for any divisor F over W_1. $\square$

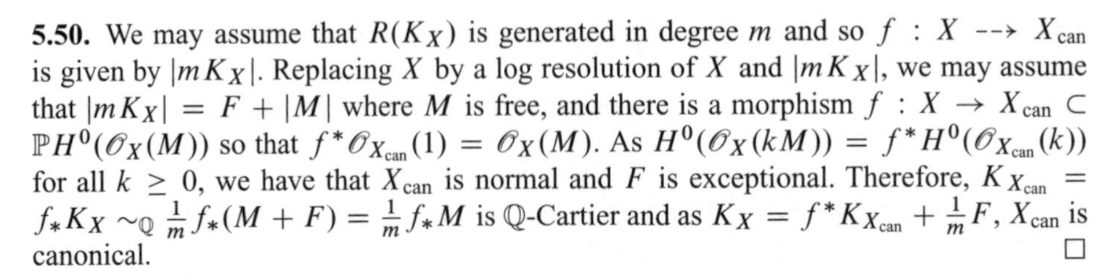

5.50. We may assume that $R(K_X)$ is generated in degree m and so $f : X \dashrightarrow X_{\mathrm{can}}$ is given by $|mK_X|$. Replacing X by a log resolution of X and $|mK_X|$, we may assume that $|mK_X| = F + |M|$ where M is free, and there is a morphism $f : X \to X_{\mathrm{can}} \subset \mathbb{P}H^0(\mathcal{O}_X(M))$ so that $f^*\mathcal{O}_{X_{\mathrm{can}}}(1) = \mathcal{O}_X(M)$. As $H^0(\mathcal{O}_X(kM)) = f^*H^0(\mathcal{O}_{X_{\mathrm{can}}}(k))$ for all $k \geq 0$, we have that X_{can} is normal and F is exceptional. Therefore, $K_{X_{\mathrm{can}}} = f_*K_X \sim_{\mathbb{Q}} \frac{1}{m} f_*(M + F) = \frac{1}{m} f_*M$ is $\mathbb{Q}$-Cartier and as $K_X = f^*K_{X_{\mathrm{can}}} + \frac{1}{m}F$, X_{can} is canonical. □

5.51. Let $\mathbb{F}_N$ be the ruled surface $\mathbb{P}_{\mathbb{P}^1}(\mathcal{O}_{\mathbb{P}^1} \oplus \mathcal{O}_{\mathbb{P}^1}(N)) \to \mathbb{P}^1$. Let X_0 be the negative section so that $X_0^2 = -N$ and let F be a fiber. We blow up a point $p \in F$ to obtain a morphism $f : X \to X_0$ with exceptional divisor E. If $\tilde{F}$ is the strict transform of F, then $\tilde{F}^2 = -1$. If $p \in X_0$ (resp. $p \notin X_0$), then $\tilde{X}_0^2 = N - 1$ (resp. $\tilde{X}_0^2 = N$). We may now contract the -1 curve $\tilde{F}$ to get a morphism $g : X \to X_{N-1}$ (resp. $g : X \to X_{N+1}$). □

5.52. Let $f : Y \to X$ be a morphism from a smooth surface and Y_{min} be the ltm of Y over X. $K_{Y_{\mathrm{min}}}$ is nef over X and Y_{min} is a terminal surface and hence Y_{min} is smooth. Suppose now that $f' : Y' \to X$ is a morphism from a smooth surface, then since $Y_{\mathrm{min}} \simeq Y'_{\mathrm{min}}$, the rational map $Y' \to Y_{\mathrm{min}}$ is a morphism. □

5.53. See [KM98, 4.11]. □

5.71. Let $\Delta = S + \sum \delta_i \Delta_i$ and set $\Delta' = S + \sum \delta_i' \Delta_i$ where $0 < \delta_i' < \delta_i$ are rational numbers such that $|\delta_i' - \delta_i| \ll 1$. It is easy to check that f is a pl-flipping contraction for (X, Δ'). □

5.72. Since Z is affine, the sheaf $f_*\mathcal{O}_X(S)$ is generated by global sections. Let $S' \sim S$ be a divisor corresponding to a general section of $\Gamma(Z, f_*\mathcal{O}_X(S)) \simeq \Gamma(X, \mathcal{O}_X(S))$. Since f is an isomorphism in codimension 1 and $\mathcal{O}_X(S)$ is invertible in codimension 1, S and S' have no common components. □

6.17. Let $f : Y \to X$ be a log resolution of (X, Δ) and V, then $M = f^*D - \mathrm{Fix}(f^*V)$ is a reduced irreducible smooth divisor intersecting $F = \mathrm{Fix}(f^*V)$ transversely. Since $c < 1$, we have

$$\mathcal{J}_{\Delta, c\cdot V} = f_*\mathcal{O}_Y(E - \lfloor cF \rfloor) = f_*\mathcal{O}_Y(E - \lfloor c(M + F) \rfloor) = \mathcal{J}_{\Delta, cD}.$$ □

6.18. Let $f : Y \to X$ be a log resolution of $(X, \Delta + D)$ that is an isomorphism at the general point of each non-klt center of (X, Δ). If $\mathcal{J}_{\Delta, D} = \mathcal{O}_X$, then $E - \lfloor f^*D \rfloor \geq 0$ and in particular $\lfloor D \rfloor = 0$. For any exceptional prime divisor F in Y, we have $a(F; X, \Delta + D) = \mathrm{mult}_F(E - f^*D) > -1$ so that $(X, \Delta + D)$ is dlt. The reverse implication is similar. □

6.19. Let $f : Y \to X$ be a log resolution of $(X, \Delta + D)$. Since $E \geq 0$, we have that $E - \lfloor cf^*D \rfloor$ is effective and f-exceptional for $0 < c \ll 1$. Therefore $\mathcal{J}_{\Delta, cD} = \mathcal{O}_X$.

If $\mathrm{mult}_x(D) > \dim(X)$, let $\mu : X' \to X$ be the blow-up of $x \in X$ with exceptional divisor F. Then $\mathrm{mult}_F(K_{X'} + (\mu^{-1})_*D - \mu^*(K_X + D)) \geq 1$. It follows that for any log resolution $f : Y \to X$ dominating μ, we have that $E - \lfloor f^*D \rfloor$ is not effective and in fact contained in $\mathcal{O}_Y(-F')$ where F' is the strict transform of F. Therefore $\mathcal{J}_{\Delta,D} \subset f_*\mathcal{O}_Y(-F') = m_x$. □

6.20. Let $p > 0$ be an integer such that $\mathcal{J}_{||D||} = \mathcal{J}_{\frac{1}{p}\cdot|pD|}$. Let $f : Y \to X$ be a log resolution of $|pD|$. Then $\mathcal{J}_{||D||} = f_*\mathcal{O}_Y\left(K_{Y/X} - \left\lfloor \frac{1}{p} F_p \right\rfloor\right)$ where $F_p = \text{Fix}(|pf^*D|)$. If $F_1 = \text{Fix}(|f^*D|)$, then $pF_1 \geq F_p$ and so

$$\begin{aligned} H^0(\mathcal{O}_X(D)) &= H^0(\mathcal{O}_Y(f^*D - F_1)) \\ &\subseteq H^0\left(\mathcal{O}_Y\left(f^*D + K_{Y/X} - \left\lfloor \frac{1}{p} F_p \right\rfloor\right)\right) = H^0(\mathcal{J}_{||D||}(D)) \subset H^0(\mathcal{O}_X(D)). \end{aligned}$$

□

6.21. For any $m > 0$, there exists a number $k > 0$ such that $\text{mult}_Z(|mkD|)/k < 1$. By [Laz04b, 9.5.13], we have that $\mathcal{J}_{\frac{1}{k}\cdot|mkD|} = \mathcal{O}_X$ on a neighborhood of a general point of Z. Let H be an ample $\mathbb{Q}$-divisor on X. For any $m > 0$ sufficiently big and divisible, $\mathcal{O}_X(m(D + H)) \otimes \mathcal{J}_{\frac{1}{k}\cdot|mkD|}$ is globally generated cf. (3) of (6.10). □

6.23. Let $f : Y \to X$ be a log resolution of $(X, \Delta + D)$. By the projection formula, we have

$$\begin{aligned} \mathcal{J}_{\Delta,D} &= f_*\mathcal{O}_Y(E - \lfloor f^*(\lfloor D \rfloor + \{D\}) \rfloor) \\ &= f_*\mathcal{O}_Y(E - \lfloor f^*\{D\} \rfloor) \otimes \mathcal{O}_X(-\lfloor D \rfloor) = \mathcal{J}_{\Delta,\{D\}} \otimes \mathcal{O}_X(-\lfloor D \rfloor). \end{aligned}$$

□

6.24. Let H be an ample divisor on X and $H' = f_*H$. Then $F = f^*H' - H$ is effective and $-F$ is ample over Y. □

6.25. Since the spectral sequence $E^1_{p,q} = \mathcal{R}^p\pi_*\mathcal{R}^q f_*F$ degenerates at the E^1-term as $E^1_{p,q} = 0$ for $q > 0$. □

6.26. See [Laz04b, 1.8.5]. □

6.28. Since $(X, f_*\Xi)$ is canonical, $K_Y + \Xi = f^*(K_X + f_*\Xi) + E$ where E is effective and exceptional. □

6.29. Let $i : S \to X$ and $j : T \to Y$ be the given inclusions. Let $D' \in |D|$ be a divisor whose support does not contain S. Then D' is locally defined by a rational function $g \in K(X)$. The corresponding section of $H^0(T, \mathcal{O}_T(f^*D))$ is defined by $(g \circ f) \circ j$ whereas the corresponding section of $H^0(T, \mathcal{O}_T((f|_T)^*D|_S))$ is defined by $(g \circ i) \circ f|_T$. Since $i \circ f|_T = f \circ j$, the excercise follows. □

8.4. If we set $T = p_*^{-1}S$, then $(p|_T)^*(K_S + \Delta_S) = (q|_T)^*(K_{S'} + \Delta_{S'}) + F|_T$. Therefore $a(E; S', \Delta_{S'}) = a(E; S, \Delta_S) + \text{mult}_E(F) \geq a(E; S, \Delta_S)$. We write $K_Y + \Gamma' = q^*(K_{X'} + \Delta')$ so that $K_Y + \Gamma' + F = p^*(K_X + \Delta)$. Then $\Delta_S = (p|_T)_*(\Gamma' + F - T)|_T$ and $\Delta_{S'} = (q|_T)_*(\Gamma' - T)|_T$. Therefore

$$(\phi|_S)_*(\Delta_S) = (q|_T)_*(\Gamma' + F - T)|_T \geq (q|_T)_*(\Gamma' - T)|_T = \Delta_{S'}.$$

□

8.5. We may assume that F is supported on a prime divisor and that $\lambda D - F \geq 0$. Let P be a prime divisor contained in the support of $\mathbf{Fix}(D)$. From the map of linear series

$$|D + F|_{\mathbb{R}} + |\lambda D - F|_{\mathbb{R}} \to |(\lambda + 1)D|_{\mathbb{R}}$$

one sees that then

$$(\lambda + 1)\operatorname{mult}_P |D|_{\mathbb{R}} = \operatorname{mult}_P |(\lambda + 1)D|_{\mathbb{R}} \leq \operatorname{mult}_P |D + F|_{\mathbb{R}} + \operatorname{mult}_P |\lambda D - F|_{\mathbb{R}}.$$

If $P \neq \operatorname{Supp}(F)$ (resp. $P = \operatorname{Supp}(F)$), then $\operatorname{mult}_P |\lambda D - F|_{\mathbb{R}} = \operatorname{mult}_P |\lambda D|_{\mathbb{R}}$ (resp. $\operatorname{mult}_P |\lambda D - F|_{\mathbb{R}} = \operatorname{mult}_P |\lambda D|_{\mathbb{R}} - \operatorname{mult}_P(F)$). Therefore $\operatorname{mult}_P |D|_{\mathbb{R}} \leq \operatorname{mult}_P |D + F|_{\mathbb{R}}$ (resp. $\operatorname{mult}_P |D|_{\mathbb{R}} + \operatorname{mult}_P(F) \leq \operatorname{mult}_P |D + F|_{\mathbb{R}}$). The assertion now follows easily. □

9.22. Let $f : Y \to X$ be a log resolution of $(X, \Delta + H)$, then since $|f^*\Delta|_{\mathbb{R}}$ is free and the coefficients of $f^*\Delta'$ are < 1, $\mathcal{J}(H) = f_*\mathcal{O}_Y(K_{Y/X} - \lfloor f^*H \rfloor) = f_*\mathcal{O}_Y(K_{Y/X} - \lfloor f^*H + f^*\Delta \rfloor) = \mathcal{J}(H + \Delta')$. □

9.23. Since D is pseudo-effective, $D \cdot H^{\dim(X)-1} = 0$. Suppose that C is a curve belonging to a covering family of X, then $D \cdot C \geq 0$. Replacing H by a multiple, we may assume that $\mathcal{O}_X(H) \otimes \mathcal{I}_C$ is generated. Therefore $H^{\dim(X)-1} \equiv C + C'$ where C' also belongs to a covering family of X. It follows that $0 = H^{\dim(X)-1} \cdot D = (C + C') \cdot D \geq 0$ and hence $C \cdot D = 0$ for any moving curve. Repeating the above argument for an arbitrary curve C, one sees that $D \equiv 0$. □

10.5. Let G be a divisor whose support contains the support of any divisor in V and let $f : Y \to X$ be a log resolution of (X, G). It follows that f is a log resolution of (X, Δ) for any $0 \leq \Delta \in V$. We let $K_Y + \Gamma = f^*(K_X + \Delta)$. (X, Δ) is klt if $\operatorname{mult}_F(\Gamma) < 1$ where F is any divisor in the support of G or any f-exceptional divisor on Y. Since (X, Δ_0) is klt and $\operatorname{mult}_F(\Gamma)$ are continuous functions of Δ, the assertion is clear. □

10.6. By the Negativity lemma, it suffices to check that
$a(F; X, \Delta) < a(F; X, \psi_*\phi_*\Delta)$ for all $(\psi \circ \phi)$-exceptional divisors $F \subset X$. Let $p : W \to X$, $q : W \to Y$ and $r : W \to Z$ be a common resolution, then we have that $q^*(K_Y + \phi_*\Delta) = r^*(K_Z + \psi_*\phi_*\Delta) + E$ where $E \geq 0$ is r-exceptional and that $K_X + \Delta = p_*q^*(K_Y + \phi_*\Delta) + G$ where $G \geq 0$ is ϕ-exceptional. Therefore $K_X + \Delta = p_*r^*(K_Z + \psi_*\phi_*\Delta) + p_*E + G$ where $p_*E + G \geq 0$. Since $\operatorname{Supp}(q_*E)$ contains all ψ-exceptional divisors and $\operatorname{Supp}(G)$ contains all ϕ-exceptional divisors, $\operatorname{Supp}(G + p_*E)$ contains all $(\phi \circ \psi)$-exceptional divisors. □

10.7. Let $\phi : X \dashrightarrow Y$ be an ltm for (X, Δ) over Z and $\zeta : Y \to Z$ be the corresponding morphism. We must show that if $||\Delta - \Delta_0|| \ll 1$ and if ϕ is an ltm for (X, Δ) over Z then ϕ is an ltm for (X, Δ) over U (the reverse implication is easier). Replacing Δ by an appropriate ($\mathbb{R}$-linearly equivalent over U) divisor, we may assume that $\Delta = A + B$ where A is ample over U. By the Cone Theorem (5.4), there are finitely many extremal rays R_i over U corresponding to curves $C_i \subset Y$ such that if $K_Y + \phi_*\Delta$ is not nef over U, then $(K_Y + \phi_*\Delta) \cdot C_i < 0$. Since $(K_Y + \phi_*\Delta) \sim_{\mathbb{R},U} \phi_*(\Delta - \Delta_0) + \zeta^*H$ and

$||\Delta - \Delta_0|| \ll 1$, we may assume that $\zeta^* H \cdot C_i = 0$ and hence that the R_i are extremal rays over Z. By assumption $K_Y + \phi_* \Delta$ is nef over Z and hence it is nef over U. □

10.8. Let Δ_i be the vertices of $\mathcal{P}$, then we may write $\Delta_i \sim_{U,\mathbb{Q}} \Delta_i' + \varepsilon A$ for all i. We may assume that $(X, \Delta_i' + \varepsilon A)$ is klt for all i. We define $L_{\mathcal{P}}$ to be a linear function on the affine subspace generated by $\mathcal{P}$ such that $L_{\mathcal{P}}(\Delta_i) = \Delta_i' + \varepsilon A$ and then pick an appropriate linear map $L : \mathrm{WDiv}_{\mathbb{R}}(X) \to \mathrm{WDiv}_{\mathbb{R}}(X)$ such that $L|_{\mathcal{P}} = L_{\mathcal{P}}$. If $\Delta \in \mathcal{P}$, then $\Delta = \sum r_i \Delta_i$ where $r_i \geq 0$ and $\sum r_i = 1$. But then

$$\phi(\Delta) = \phi(\sum r_i \Delta_i) = \sum r_i \phi(\Delta_i) = \sum r_i (\Delta_i' + \varepsilon A) \sim_{U,\mathbb{Q}} \sum r_i (\Delta_i) = \Delta.$$

Let F be any divisor over X, since $(X, \Delta_i' + \varepsilon A)$ is klt, $a(F; X, \Delta_i' + \varepsilon A) > -1$. It follows that

$$a(F; X, \phi(\Delta)) = a(F; X, \sum r_i (\Delta_i' + \varepsilon A)) = \sum r_i a(F; X, \Delta_i' + \varepsilon A) > -\sum r_i = -1.$$

It is easy to see that $L(\Delta) \geq \varepsilon A$. □

11.2. See [Har77, Ex. III.12.4]. □

12.7. Let Y be Hironaka's example for a non-singular proper variety that is not projective explained in [Har77, B.3.4.1]. Then from the construction of Y it follows that it admits a morphism $g : Y \to T$ to a non-singular projective threefold T. Since both Y and T are proper, the morphism g is also proper. However, if it were projective, then Y would also be projective. Again from the construction we obtain that g is birational and the fibers where it is not an isomorphism are either a smooth projective rational curve or the union of two smooth projective rational curves intersecting transversally in a single point. All of these are clearly projective varieties. □

12.9. The only varieties with Hilbert polynomial 1 are single points and hence the points of $\mathrm{Hilb}_1(X/S)$ correspond to the points of X/S. □

16.1. The Satake compactification provides an embedding $\mathsf{M}_g \hookrightarrow M$ where M is a projective variety and $\mathrm{codim}_M(M \setminus \mathsf{M}_g) = 2$. Let $\sigma : \widetilde{M} \to M$ be the blow-up of M along $x_1, \dots, x_r$ and let $E_i = \sigma^{-1}(x_i)$ for $i = 1, \dots, r$. Further let H be a very ample Cartier divisor on M. Then possibly replacing H with a multiple of itself, one may assume that $\widetilde{H} = \sigma^* H - \sum_{i=1}^r E_i$ is a very ample divisor cf. [KM98, 2.62]. Let $\widetilde{C} \subset \widetilde{M}$ be a general complete intersection curve corresponding to the complete linear system of $\widetilde{H}$. Since the x_i are disjoint from $M \setminus \mathsf{M}_g$, it follows that $\widetilde{M} \setminus \sigma^{-1}(\mathsf{M}_g) \simeq M \setminus \mathsf{M}_g$, in particular it has codimension 2. It follows that $\widetilde{C} \subset \sigma^{-1}(\mathsf{M}_g)$ and that $\widetilde{C} \cap E_i \neq \emptyset$. Therefore, $C = \sigma(\widetilde{C})$ satisfies the requirements. □

16.2. Let $g \geq 3$ and $f \in H^0(\mathsf{M}_g, \mathscr{O}_{\mathsf{M}_g})$ a global regular function. Let $x, y \in \mathsf{M}_g$ be two arbitrary general points. Then by (16.1) there exists a proper curve $C \subset \mathsf{M}_g$ such that $x, y \in C$. Since $f\big|_C$ has to be constant, $f(x) = f(y)$. Therefore f is constant on general points, but then it has to be constant everywhere.

This obviously implies that M_g is not affine.

The Satake compactification provides an embedding $\mathsf{M}_g \hookrightarrow M$ where M is a projective variety and $\operatorname{codim}_M(M \setminus \mathsf{M}_g) = 2$. This proves that M_g is not projective. (This could be proved in many other ways.) $\square$

16.6. If $g(B) \geq 2$, then $2g(B) - 2 + \#\Delta > 0$ regardless of Δ. If $g(B) = 1$, then $2g(B) - 2 + \#\Delta = \#\Delta$, so it is non-positive if and only if $\Delta = \emptyset$. If $g(B) = 0$, then $2g(B) - 2 + \#\Delta = \#\Delta - 2$, so it is non-positive if and only if $\#\Delta \leq 2$. $\square$

16.25. There exist dominant holomorphic maps $\mathbb{C}^\times \hookrightarrow \mathbb{C}$ (the usual inclusion) and $\mathbb{C} \twoheadrightarrow \mathbb{C}^\times$ (the exponential map). Therefore there exists a non-constant holomorphic map $\mathbb{C} \to X$ if and only if there exists a non-constant holomorphic map $\mathbb{C}^\times \to X$. $\square$

16.26. If T is a complex torus of dimension n, then there exists a surjective holomorphic map $\lambda : \mathbb{C}^n \twoheadrightarrow T$. If $\tau : T \to X$ is a non-constant holomorphic map, then $\tau \circ \lambda$ is a non-constant holomorphic map and hence X cannot be Brody hyperbolic. $\square$

Bibliography

[AH09] D. ABRAMOVICH AND B. HASSETT: *Stable varieties with a twist*, to appear in the proceedings of the conference 'Classification of algebraic varieties' at Schiermonnikoog, the Netherlands. arXiv:0904.2797v1 [math.AG]

[AN54] Y. AKIZUKI AND S. NAKANO: *Note on Kodaira-Spencer's proof of Lefschetz theorems*, Proc. Japan Acad. **30** (1954), 266–272. MR0066694 (16,619a)

[Ale94] V. ALEXEEV: *Boundedness and K^2 for log surfaces*, Internat. J. Math. **5** (1994), no. 6, 779–810. MR1298994 (95k:14048)

[Ale96] V. ALEXEEV: *Moduli spaces $M_{g,n}(W)$ for surfaces*, Higher-dimensional complex varieties (Trento, 1994), de Gruyter, Berlin, 1996, pp. 1–22. MR1463171 (99b:14010)

[Ale01] V. ALEXEEV: *On extra components in the functorial compactification of A_g*, Moduli of abelian varieties (Texel Island, 1999), Progr. Math., vol. 195, Birkhäuser, Basel, 2001, pp. 1–9. MR1827015 (2002d:14070)

[Ale02] V. ALEXEEV: *Complete moduli in the presence of semiabelian group action*, Ann. of Math. (2) **155** (2002), no. 3, 611–708. MR1923963 (2003g:14059)

[AB04a] V. ALEXEEV AND M. BRION: *Stable reductive varieties. I. Affine varieties*, Invent. Math. **157** (2004), no. 2, 227–274. MR2076923

[AB04b] V. ALEXEEV AND M. BRION: *Stable reductive varieties. II. Projective case*, Adv. Math. **184** (2004), no. 2, 380–408. MR2054021

[AB05] V. ALEXEEV AND M. BRION: *Moduli of affine schemes with reductive group action*, J. Algebraic Geom. **14** (2005), no. 1, 83–117. MR2092127

[AHK07] V. ALEXEEV, C. HACON, AND Y. KAWAMATA: *Termination of (many) 4-dimensional log flips*, Invent. Math. **168** (2007), no. 2, 433–448. MR2289869 (2008f:14028)

[AM04] V. ALEXEEV AND S. MORI: *Bounding singular surfaces of general type*, Algebra, arithmetic and geometry with applications (West Lafayette, IN, 2000), Springer, Berlin, 2004, pp. 143–174. MR2037085

[AB04] V. A. ALEXEEV AND M. BRION: *Boundedness of spherical Fano varieties*, The Fano Conference, Univ. Torino, Turin, 2004, pp. 69–80. MR2112568 (2005k:14081)

[Amb03] F. AMBRO: *Quasi-log varieties*, Tr. Mat. Inst. Steklova **240** (2003), no. Biratsion. Geom. Linein. Sist. Konechno Porozhdennye Algebry, 220–239. MR1993751 (2004f:14027)

[Amb06] F. AMBRO: *Restrictions of log canonical algebras of general type*, J. Math. Sci. Univ. Tokyo **13** (2006), no. 3, 409–437. MR2284409 (2007m:14015)

[Ara71] S. J. ARAKELOV: *Families of algebraic curves with fixed degeneracies*, Izv. Akad. Nauk SSSR Ser. Mat. **35** (1971), 1269–1293. MR0321933 (48 #298)

[Băd01] L. BĂDESCU: *Algebraic surfaces*, Universitext, Springer-Verlag, New York, 2001, Translated from the 1981 Romanian original by Vladimir Masek and revised by the author. MR1805816 (2001k:14068)

[BB66] W. L. BAILY, JR. AND A. BOREL: *Compactification of arithmetic quotients of bounded symmetric domains*, Ann. of Math. (2) **84** (1966), 442–528. MR0216035 (35 #6870)

[BHPV04] W. P. BARTH, K. HULEK, C. A. M. PETERS, AND A. VAN DE VEN: *Compact complex surfaces*, second ed., Ergebnisse der Mathematik und ihrer Grenzgebiete. 3. Folge. A Series of Modern Surveys in Mathematics [Results in Mathematics and Related Areas. 3rd Series], vol. 4, Springer-Verlag, Berlin, 2004. MR2030225 (2004m:14070)

[Bea96] A. BEAUVILLE: *Complex algebraic surfaces*, second ed., London Mathematical Society Student Texts, vol. 34, Cambridge University Press, Cambridge, 1996, Translated from the 1978 French original by R. Barlow, with assistance from N. I. Shepherd-Barron and M. Reid. 97e:14045

[BV00] E. BEDULEV AND E. VIEHWEG: *On the Shafarevich conjecture for surfaces of general type over function fields*, Invent. Math. **139** (2000), no. 3, 603–615. MR1738062 (2001f:14065)

[BP08] B. BERNDTSSON AND M. PĂUN: *A Bergman kernel proof of the Kawamata subadjunction theorem*, 2008.

[Bir03] C. BIRKAR: *On Shokurov's log flips: The 3-dimensional case*, 2003.

[BCHM10] C. BIRKAR, P. CASCINI, C. D. HACON, AND J. MCKERNAN: *Existence of minimal models for varieties of log general type*, Journal of the AMS **23** (2010), 405–468. math.AG/0610203v2, DOI:10.1090/S0894-0347-09-00649-3

[BB92] A. A. BORISOV AND L. A. BORISOV: *Singular toric Fano three-folds*, Mat. Sb. **183** (1992), no. 2, 134–141. MR1166957 (93i:14034)

[Bor01] A. BORISOV: *Boundedness of Fano threefolds with log-terminal singularities of given index*, J. Math. Sci. Univ. Tokyo **8** (2001), no. 2, 329–342. MR1837167 (2002d:14060)

[BLR90] S. BOSCH, W. LÜTKEBOHMERT, AND M. RAYNAUD: *Néron models*, Ergebnisse der Mathematik und ihrer Grenzgebiete (3) [Results in Mathematics and Related Areas (3)], vol. 21, Springer-Verlag, Berlin, 1990. MR1045822 (91i:14034)

[BDPP03] S. BOUCKSOM, J.-P. DEMAILLY, M. PĂUN, AND T. PETERNELL: *The pseudo-effecitve cone of a compact Kähler manifold and varieties of negative Kodaira dimension*, preprint, 2003.

[Bro78] R. BRODY: *Compact manifolds in hyperbolicity*, Trans. Amer. Math. Soc. **235** (1978), 213–219. MR0470252 (57 #10010)

[BH93] W. BRUNS AND J. HERZOG: *Cohen-Macaulay rings*, Cambridge Studies in Advanced Mathematics, vol. 39, Cambridge University Press, Cambridge, 1993. MR1251956 (95h:13020)

[Cap02] L. CAPORASO: *On certain uniformity properties of curves over function fields*, Compositio Math. **130** (2002), no. 1, 1–19. MR1883689 (2003a:14038)

[Cap03] L. CAPORASO: *Remarks about uniform boundedness of rational points over function fields*, Trans. Amer. Math. Soc. **355** (2003), no. 9, 3475–3484 (electronic). MR1990159 (2004d:14019)

[Cap04] L. CAPORASO: *Moduli theory and arithmetic of algebraic varieties*, The Fano Conference, Univ. Torino, Turin, 2004, pp. 353–378. MR2112582

[Car85] J. A. CARLSON: *Polyhedral resolutions of algebraic varieties*, Trans. Amer. Math. Soc. **292** (1985), no. 2, 595–612. MR808740 (87i:14008)

[Che04] X. CHEN: *On algebraic hyperbolicity of log varieties*, Commun. Contemp. Math. **6** (2004), no. 4, 513–559. MR2078413 (2005k:14089)

[CKM88] H. CLEMENS, J. KOLLÁR, AND S. MORI: *Higher-dimensional complex geometry*, Astérisque (1988), no. 166, 144 pp. (1989). MR1004926 (90j:14046)

[Cor07] A. CORTI: *3-fold flips after Shokurov*, Flips for 3-folds and 4-folds, Oxford Lecture Ser. Math. Appl., vol. 35, Oxford Univ. Press, Oxford, 2007, pp. 18–48. MR2359340

[CHK+07] A. CORTI, P. HACKING, J. KOLLÁR, R. LAZARSFELD, AND M. MUSTAŢĂ: *Lectures on flips and minimal models*, 2007.

[CKL08] A. CORTI, A.-S. KALOGHIROS, AND V. LAZIC: *Introduction to the minimal model program and the existence of flips*, 2008.

[dF13] M. DE FRANCHIS: *Un teorema sulle involuzioni irrazionali*, Rend. Circ. Mat. Palermo (1913), no. 36, 368.

[dF91] M. DE FRANCHIS: *Collected works of Michele de Franchis*, Rend. Circ. Mat. Palermo (2) Suppl. (1991), no. 27, vii+636, Edited by C. Ciliberto and E. Sernesi. MR1174233 (93i:01031)

[Deb01] O. DEBARRE: *Higher-dimensional algebraic geometry*, Universitext, Springer-Verlag, New York, 2001. MR1841091 (2002g:14001)

[Dem97] J.-P. DEMAILLY: *Algebraic criteria for Kobayashi hyperbolic projective varieties and jet differentials*, Algebraic geometry—Santa Cruz 1995, Proc. Sympos. Pure Math., vol. 62, Amer. Math. Soc., Providence, RI, 1997, pp. 285–360. MR1492539 (99b:32037)

[Dia84] S. DIAZ: *A bound on the dimensions of complete subvarieties of* $\mathscr{M}_g$, Duke Math. J. **51** (1984), no. 2, 405–408. MR747872 (85j:14042)

[Dia87] S. DIAZ: *Complete subvarieties of the moduli space of smooth curves*, Algebraic geometry, Bowdoin, 1985 (Brunswick, Maine, 1985), Proc. Sympos. Pure Math., vol. 46, Amer. Math. Soc., Providence, RI, 1987, pp. 77–81. MR927950 (89b:14039)

[DB81] P. DU BOIS: *Complexe de de Rham filtré d'une variété singulière*, Bull. Soc. Math. France **109** (1981), no. 1, 41–81. MR613848 (82j:14006)

[DJ74] P. DUBOIS AND P. JARRAUD: *Une propriété de commutation au changement de base des images directes supérieures du faisceau structural*, C. R. Acad. Sci. Paris Sér. A **279** (1974), 745–747. MR0376678 (51 #12853)

[Dur79] A. H. DURFEE: *Fifteen characterizations of rational double points and simple critical points*, Enseign. Math. (2) **25** (1979), no. 1-2, 131–163. MR543555 (80m:14003)

[Ein97] L. EIN: *Multiplier ideals, vanishing theorems and applications*, Algebraic geometry—Santa Cruz 1995, Proc. Sympos. Pure Math., vol. 62, Amer. Math. Soc., Providence, RI, 1997, pp. 203–219. MR1492524 (98m:14006)

[Eis95] D. EISENBUD: *Commutative algebra*, Graduate Texts in Mathematics, vol. 150, Springer-Verlag, New York, 1995, With a view toward algebraic geometry. MR1322960 (97a:13001)

[EH00] D. EISENBUD AND J. HARRIS: *The geometry of schemes*, Graduate Texts in Mathematics, vol. 197, Springer-Verlag, New York, 2000. MR1730819 (2001d:14002)

[Elk78] R. ELKIK: *Singularités rationnelles et déformations*, Invent. Math. **47** (1978), no. 2, 139–147. MR501926 (80c:14004)

[Elk81] R. ELKIK: *Rationalité des singularités canoniques*, Invent. Math. **64** (1981), no. 1, 1–6. MR621766 (83a:14003)

[EV90] H. Esnault and E. Viehweg: *Effective bounds for semipositive sheaves and for the height of points on curves over complex function fields*, Compositio Math. **76** (1990), no. 1-2, 69–85, Algebraic geometry (Berlin, 1988). MR1078858 (91m:14038)

[EV92] H. Esnault and E. Viehweg: *Lectures on vanishing theorems*, DMV Seminar, vol. 20, Birkhäuser Verlag, Basel, 1992. MR1193913 (94a:14017)

[FL99] C. Faber and E. Looijenga: *Remarks on moduli of curves*, Moduli of curves and abelian varieties, Aspects Math., E33, Vieweg, Braunschweig, 1999, pp. 23–45. MR1722537 (2001b:14044)

[FvdG04] C. Faber and G. van der Geer: *Complete subvarieties of moduli spaces and the Prym map*, J. Reine Angew. Math. **573** (2004), 117–137. MR2084584 (2005g:14054)

[Fal83a] G. Faltings: *Arakelov's theorem for abelian varieties*, Invent. Math. **73** (1983), no. 3, 337–347. MR718934 (85m:14061)

[Fal83b] G. Faltings: *Endlichkeitssätze für abelsche Varietäten über Zahlkörpern*, Invent. Math. **73** (1983), no. 3, 349–366. MR718935 (85g:11026a)

[Fal84] G. Faltings: *Erratum: "Finiteness theorems for abelian varieties over number fields"*, Invent. Math. **75** (1984), no. 2, 381. MR732554 (85g:11026b)

[Fon85] J.-M. Fontaine: *Il n'y a pas de variété abélienne sur* **Z**, Invent. Math. **81** (1985), no. 3, 515–538. MR807070 (87g:11073)

[Fuj05] O. Fujino: *Addendum to: "Termination of 4-fold canonical flips" [Publ. Res. Inst. Math. Sci.* **40** *(2004), no. 1, 231–237; mr2030075]*, Publ. Res. Inst. Math. Sci. **41** (2005), no. 1, 251–257. MR2115973 (2005j:14013)

[Fuj07] O. Fujino: *Special termination and reduction to pl flips*, Flips for 3-folds and 4-folds, Oxford Lecture Ser. Math. Appl., vol. 35, Oxford Univ. Press, Oxford, 2007, pp. 63–75. MR2359342

[Fuj08a] O. Fujino: *Introduction to the log minimal model program for log canonical pairs*, unpublished manuscript, 2008.

[Fuj08b] O. Fujino: *Theory of non-lc ideal sheaves–basic properties–*, arXiv:0801.2198

[FM00] O. Fujino and S. Mori: *A canonical bundle formula*, J. Differential Geom. **56** (2000), no. 1, 167–188. MR1863025 (2002h:14091)

[Fuj78] T. Fujita: *On Kähler fiber spaces over curves*, J. Math. Soc. Japan **30** (1978), no. 4, 779–794. MR513085 (82h:32024)

[GDH99] G. González-Díez and W. J. Harvey: *Subvarieties of moduli space for Riemann surfaces*, Complex geometry of groups (Olmué, 1998), Contemp. Math., vol. 240, Amer. Math. Soc., Providence, RI, 1999, pp. 197–208. MR1703560 (2001a:32019)

[GR70] H. Grauert and O. Riemenschneider: *Verschwindungssätze für analytische Kohomologiegruppen auf komplexen Räumen*, Invent. Math. **11** (1970), 263–292. MR0302938 (46 #2081)

[GKK10] D. Greb, S. Kebekus, and S. J. Kovács: *Extension theorems for differential forms, and Bogomolov-Sommese vanishing on log canonical varieties*, Compos. Math. **146** (2010), 193–219. Published online by Cambridge University Press Dec. 11, 2009. DOI: 10.1112/S0010437X09004321

[GKKP10] D. Greb, S. Kebekus, S. J. Kovács, and T. Peternell: *Differential forms on log canonical spaces*, in preparation, 2010.

[Gro62] A. Grothendieck: *Fondements de la géométrie algébrique. [Extraits du Séminaire Bourbaki, 1957–1962.]*, Secrétariat mathématique, Paris, 1962. MR0146040 (26 #3566)

[Gro95] A. Grothendieck: *Fondements de la géométrie algébrique. Commentaires [MR0146040 (26 #3566)]*, Séminaire Bourbaki, Vol. 7, Soc. Math. France, Paris, 1995, pp. 297–307. MR1611235

[GNPP88] F. Guillén, V. Navarro Aznar, P. Pascual Gainza, and F. Puerta: *Hyperrésolutions cubiques et descente cohomologique*, Lecture Notes in Mathematics, vol. 1335, Springer-Verlag, Berlin, 1988, Papers from the Seminar on Hodge-Deligne Theory held in Barcelona, 1982. MR972983 (90a:14024)

[Hac04] P. Hacking: *Compact moduli of plane curves*, Duke Math. J. **124** (2004), no. 2, 213–257. MR2078368 (2005f:14056)

[HKT06] P. Hacking, S. Keel, and J. Tevelev: *Compactification of the moduli space of hyperplane arrangements*, J. Algebraic Geom. **15** (2006), no. 4, 657–680. MR2237265 (2007j:14016)

[HKT07] P. Hacking, S. Keel, and J. Tevelev: *Stable pair, tropical, and log canonical compact moduli of del Pezzo surfaces*, 2007. arXiv:math/0702505

[HdF08] C. D. Hacon and T. de Fernex: *Singularities on normal varieties*, 2008. arXiv:0805.1767v1 [math.AG]

[HM06] C. D. Hacon and J. McKernan: *Boundedness of pluricanonical maps of varieties of general type*, Invent. Math. **166** (2006), no. 1, 1–25. MR2242631 (2007e:14022)

[HM07] C. D. Hacon and J. McKernan: *Extension theorems and the existence of flips*, Flips for 3-folds and 4-folds, Oxford Lecture Ser. Math. Appl., vol. 35, Oxford Univ. Press, Oxford, 2007, pp. 76–110. MR2359343

[HM08] C. D. Hacon and J. McKernan: *Existence of minimal models for varieties of log general type II*, preprint, 2008. arXiv:math.AG/0808.1929

[Harr95] J. HARRIS: *Algebraic geometry*, Graduate Texts in Mathematics, vol. 133, Springer-Verlag, New York, 1995, A first course, Corrected reprint of the 1992 original. MR1416564 (97e:14001)

[HMo98] J. HARRIS AND I. MORRISON: *Moduli of curves*, Graduate Texts in Mathematics, vol. 187, Springer-Verlag, New York, 1998. MR1631825 (99g:14031)

[Har66] R. HARTSHORNE: *Residues and duality*, Lecture notes of a seminar on the work of A. Grothendieck, given at Harvard 1963/64. With an appendix by P. Deligne. Lecture Notes in Mathematics, No. 20, Springer-Verlag, Berlin, 1966. MR0222093 (36 #5145)

[Har77] R. HARTSHORNE: *Algebraic geometry*, Springer-Verlag, New York, 1977, Graduate Texts in Mathematics, No. 52. MR0463157 (57 #3116)

[HK04] B. HASSETT AND S. J. KOVÁCS: *Reflexive pull-backs and base extension*, J. Algebraic Geom. **13** (2004), no. 2, 233–247. MR2047697 (2005b:14028)

[Hei04] G. HEIER: *Uniformly effective Shafarevich conjecture on families of hyperbolic curves over a curve with prescribed degeneracy locus*, J. Math. Pures Appl. (9) **83** (2004), no. 7, 845–867. MR2074680 (2005e:14043)

[Hei09] G. HEIER: *Uniformly effective boundedness of Shafarevich conjecture-type for families of canonically polarized manifolds*, preprint, 2009. arXiv:0910.0815v1 [math.AG]

[Hir64] H. HIRONAKA: *Resolution of singularities of an algebraic variety over a field of characteristic zero. I, II*, Ann. of Math. (2) 79 (1964), 109–203; ibid. (2) **79** (1964), 205–326. MR0199184 (33 #7333)

[Iit82] S. IITAKA: *Algebraic geometry*, Graduate Texts in Mathematics, vol. 76, Springer-Verlag, New York, 1982, An introduction to birational geometry of algebraic varieties, North-Holland Mathematical Library, 24. 84j:14001

[Ish85] S. ISHII: *On isolated Gorenstein singularities*, Math. Ann. **270** (1985), no. 4, 541–554. MR776171 (86j:32024)

[Ish86] S. ISHII: *Small deformations of normal singularities*, Math. Ann. **275** (1986), no. 1, 139–148. MR849059 (87i:14003)

[Ish87a] S. ISHII: *Du Bois singularities on a normal surface*, Complex analytic singularities, Adv. Stud. Pure Math., vol. 8, North-Holland, Amsterdam, 1987, pp. 153–163. MR894291 (88f:14033)

[Ish87b] S. ISHII: *Isolated Q-Gorenstein singularities of dimension three*, Complex analytic singularities, Adv. Stud. Pure Math., vol. 8, North-Holland, Amsterdam, 1987, pp. 165–198. MR894292 (89d:32016)

[JT02] J. JORGENSON AND A. TODOROV: *Ample divisors, automorphic forms and Shafarevich's conjecture*, Mirror symmetry, IV (Montreal, QC, 2000), AMS/IP Stud. Adv. Math., vol. 33, Amer. Math. Soc., Providence, RI, 2002, pp. 361–381. MR1969038 (2004g:14036)

[Kar00] K. KARU: *Minimal models and boundedness of stable varieties*, J. Algebraic Geom. **9** (2000), no. 1, 93–109. MR1713521 (2001g:14059)

[Kas68] A. KAS: *On deformations of a certain type of irregular algebraic surface*, Amer. J. Math. **90** (1968), 789–804. MR0242199 (39 #3532)

[Kaw07] M. KAWAKITA: *Inversion of adjunction on log canonicity*, Invent. Math. **167** (2007), no. 1, 129–133. MR2264806 (2008a:14025)

[Kaw82a] Y. KAWAMATA: *A generalization of Kodaira-Ramanujam's vanishing theorem*, Math. Ann. **261** (1982), no. 1, 43–46. MR675204 (84i:14022)

[Kaw82b] Y. KAWAMATA: *Kodaira dimension of algebraic fiber spaces over curves*, Invent. Math. **66** (1982), no. 1, 57–71. MR652646 (83h:14025)

[Kaw84] Y. KAWAMATA: *The cone of curves of algebraic varieties*, Ann. of Math. (2) **119** (1984), no. 3, 603–633. MR744865 (86c:14013b)

[Kaw86] Y. KAWAMATA: *On the plurigenera of minimal algebraic 3-folds with $K \equiv 0$.*, Math. Ann. **275** (1986), no. 4, 539–546. MR859328 (88c:14049)

[Kaw92a] Y. KAWAMATA: *Abundance theorem for minimal threefolds*, Invent. Math. **108** (1992), no. 2, 229–246. MR1161091 (93f:14012)

[Kaw92b] Y. KAWAMATA: *Boundedness of $\mathbb{Q}$-Fano threefolds*, Proceedings of the International Conference on Algebra, Part 3 (Novosibirsk, 1989) (Providence, RI), Contemp. Math., vol. 131, Amer. Math. Soc., 1992, pp. 439–445. MR1175897 (93g:14047)

[Kaw99a] Y. KAWAMATA: *Deformations of canonical singularities*, J. Amer. Math. Soc. **12** (1999), no. 1, 85–92. MR1631527 (99g:14003)

[Kaw99b] Y. KAWAMATA: *On the extension problem of pluricanonical forms*, Algebraic geometry: Hirzebruch 70 (Warsaw, 1998), Contemp. Math., vol. 241, Amer. Math. Soc., Providence, RI, 1999, pp. 193–207. MR1718145 (2000i:14053)

[Kaw08] Y. KAWAMATA: *Finite generation of a canonical ring*, 2008.

[KMM87] Y. KAWAMATA, K. MATSUDA, AND K. MATSUKI: *Introduction to the minimal model problem*, Algebraic geometry, Sendai, 1985, Adv. Stud. Pure Math., vol. 10, North-Holland, Amsterdam, 1987, pp. 283–360. MR946243 (89e:14015)

[KK07] S. KEBEKUS AND S. J. KOVÁCS: *The structure of surfaces mapping to the moduli stack of canonically polarized varieties*, preprint, July 2007. arXiv:0707.2054v1 [math.AG]

[KK08a] S. KEBEKUS AND S. J. KOVÁCS: *Families of canonically polarized varieties over surfaces*, Invent. Math. **172** (2008), no. 3, 657–682. DOI: 10.1007/s00222-008-0128-8

[KK08b] S. KEBEKUS AND S. J. KOVÁCS: *Families of varieties of general type over compact bases*, Advances in Mathematics **218** (2008), 649–652. doi:10.1016/j.aim.2008.01.005

[KK08c] S. KEBEKUS AND S. J. KOVÁCS: *The structure of surfaces and threefolds mapping to the moduli stack of canonically polarized varieties*, to appear in Duke Math. J. arXiv:0812.2305v1 [math.AG]

[KMMc94] S. KEEL, K. MATSUKI, AND J. MCKERNAN: *Log abundance theorem for threefolds*, Duke Math. J. **75** (1994), no. 1, 99–119. MR1284817 (95g:14021)

[KeM97] S. KEEL AND S. MORI: *Quotients by groupoids*, Ann. of Math. (2) **145** (1997), no. 1, 193–213. MR1432041 (97m:14014)

[KS03] S. KEEL AND L. SADUN: *Oort's conjecture for $A_g \otimes \mathbb{C}$*, J. Amer. Math. Soc. **16** (2003), no. 4, 887–900 (electronic). MR1992828 (2004d:14064)

[KKMSD73] G. KEMPF, F. F. KNUDSEN, D. MUMFORD, AND B. SAINT-DONAT: *Toroidal embeddings. I*, Springer-Verlag, Berlin, 1973, Lecture Notes in Mathematics, Vol. 339. MR0335518 (49 #299)

[Kob70] S. KOBAYASHI: *Hyperbolic manifolds and holomorphic mappings*, Pure and Applied Mathematics, vol. 2, Marcel Dekker Inc., New York, 1970. MR0277770 (43 #3503)

[Kod53] K. KODAIRA: *On a differential-geometric method in the theory of analytic stacks*, Proc. Nat. Acad. Sci. U. S. A. **39** (1953), 1268–1273. MR0066693 (16,618b)

[Kod67] K. KODAIRA: *A certain type of irregular algebraic surfaces*, J. Analyse Math. **19** (1967), 207–215. MR0216521 (35 #7354)

[Kol84] J. KOLLÁR: *The cone theorem. Note to a paper: "The cone of curves of algebraic varieties"* [Ann. of Math. (2) **119** (1984), no. 3, 603–633; MR0744865 (86c:14013b)] *by Y. Kawamata*, Ann. of Math. (2) **120** (1984), no. 1, 1–5. MR750714 (86c:14013c)

[Kol85] J. KOLLÁR: *Toward moduli of singular varieties*, Compositio Math. **56** (1985), no. 3, 369–398. MR814554 (87e:14009)

[Kol86] J. KOLLÁR: *Higher direct images of dualizing sheaves. I*, Ann. of Math. (2) **123** (1986), no. 1, 11–42. MR825838 (87c:14038)

[Kol87a] J. KOLLÁR: *Subadditivity of the Kodaira dimension: fibers of general type*, Algebraic geometry, Sendai, 1985, Adv. Stud. Pure Math., vol. 10, North-Holland, Amsterdam, 1987, pp. 361–398. MR946244 (89i:14029)

[Kol87b] J. KOLLÁR: *Subadditivity of the Kodaira dimension: fibers of general type*, Algebraic geometry, Sendai, 1985, Adv. Stud. Pure Math., vol. 10, North-Holland, Amsterdam, 1987, pp. 361–398. MR946244 (89i:14029)

[Kol87c] J. Kollár: *Vanishing theorems for cohomology groups*, Algebraic geometry, Bowdoin, 1985 (Brunswick, Maine, 1985), Proc. Sympos. Pure Math., vol. 46, Amer. Math. Soc., Providence, RI, 1987, pp. 233–243. MR927959 (89j:32039)

[Kol90a] J. Kollár: *Projectivity of complete moduli*, J. Differential Geom. **32** (1990), no. 1, 235–268. MR1064874 (92e:14008)

[Kol90b] J. Kollár: *Projectivity of complete moduli*, J. Differential Geom. **32** (1990), no. 1, 235–268. MR1064874 (92e:14008)

[Kol91] J. Kollár: *Flips, flops, minimal models, etc*, Surveys in differential geometry (Cambridge, MA, 1990), Lehigh Univ., Bethlehem, PA, 1991, pp. 113–199. MR1144527 (93b:14059)

[Kol93] J. Kollár: *Effective base point freeness*, Math. Ann. **296** (1993), no. 4, 595–605. MR1233485 (94f:14004)

[Kol94] J. Kollár: *Moduli of polarized schemes*, unpublished manuscript, 1994.

[Kol95] J. Kollár: *Shafarevich maps and automorphic forms*, M. B. Porter Lectures, Princeton University Press, Princeton, NJ, 1995. MR1341589 (96i:14016)

[Kol96] J. Kollár: *Rational curves on algebraic varieties*, Ergebnisse der Mathematik und ihrer Grenzgebiete. 3. Folge. A Series of Modern Surveys in Mathematics, vol. 32, Springer-Verlag, Berlin, 1996. MR1440180 (98c:14001)

[Kol97a] J. Kollár: *Quotient spaces modulo algebraic groups*, Ann. of Math. (2) **145** (1997), no. 1, 33–79. MR1432036 (97m:14013)

[Kol97b] J. Kollár: *Singularities of pairs*, Algebraic geometry—Santa Cruz 1995, Proc. Sympos. Pure Math., vol. 62, Amer. Math. Soc., Providence, RI, 1997, pp. 221–287. MR1492525 (99m:14033)

[Kol07a] J. Kollár: *Kodaira's canonical bundle formula and adjunction*, Flips for 3-folds and 4-folds, Oxford Lecture Ser. Math. Appl., vol. 35, Oxford Univ. Press, Oxford, 2007, pp. 134–162. MR2359346

[Kol07b] J. Kollár: *Lectures on resolution of singularities*, Annals of Mathematics Studies, vol. 166, Princeton University Press, Princeton, NJ, 2007. MR2289519 (2008f:14026)

[Kol07c] J. Kollár: *Two examples of surfaces with normal crossing singularities*, 2007. arXiv:0705.0926v2

[Kol08a] J. Kollár: *Hulls and husks*, 2008. arXiv:0805.0576v2 [math.AG]

[Kol08b] J. Kollár: *Quotients by finite equivalence relations*, 2008. arXiv:0812.3608 [math.AG]

[Kol08c] J. Kollár: *Semi log resolutions*, 2008. arXiv:0812.3592v1 [math.AG]

[Kol09] J. KOLLÁR: *Simultaneous normalization and algebra husks*, 2009. arXiv:0910.1076v2 [math.AG]

[KK] J. KOLLÁR AND S. J. KOVÁCS: *Birational geometry of log surfaces*, unpublished.

[KK10] J. KOLLÁR AND S. J. KOVÁCS: *Log canonical singularities are Du Bois*, Journal of the AMS (2010), 23 pages, published eclectronically. DOI: 10.1090/S0894-0347-10-00663-6

[KMMT00] J. KOLLÁR, Y. MIYAOKA, S. MORI, AND H. TAKAGI: *Boundedness of canonical $\mathbb{Q}$-Fano 3-folds*, Proc. Japan Acad. Ser. A Math. Sci. **76** (2000), no. 5, 73–77. MR1771144 (2001h:14053)

[KM98] J. KOLLÁR AND S. MORI: *Birational geometry of algebraic varieties*, Cambridge Tracts in Mathematics, vol. 134, Cambridge University Press, Cambridge, 1998, With the collaboration of C. H. Clemens and A. Corti, Translated from the 1998 Japanese original. MR1658959 (2000b:14018)

[KSB88] J. KOLLÁR AND N. I. SHEPHERD-BARRON: *Threefolds and deformations of surface singularities*, Invent. Math. **91** (1988), no. 2, 299–338. MR922803 (88m:14022)

[Kol] J. KOLLÁR ET AL.: *Compact moduli spaces of stable varieties*, in preparation.

[Kol92] J. KOLLÁR ET AL.: *Flips and abundance for algebraic threefolds*, Société Mathématique de France, Paris, 1992, Papers from the Second Summer Seminar on Algebraic Geometry held at the University of Utah, Salt Lake City, Utah, August 1991, Astérisque No. 211 (1992). MR1225842 (94f:14013)

[Kov96] S. J. KOVÁCS: *Smooth families over rational and elliptic curves*, J. Algebraic Geom. **5** (1996), no. 2, 369–385, Erratum: J. Algebraic Geom. **6** (1997), no. 2, 391. MR1374712 (97c:14035)

[Kov97a] S. J. KOVÁCS: *Families over a base with a birationally nef tangent bundle*, Math. Ann. **308** (1997), no. 2, 347–359. MR1464907 (98h:14039)

[Kov97b] S. J. KOVÁCS: *On the minimal number of singular fibres in a family of surfaces of general type*, J. Reine Angew. Math. **487** (1997), 171–177. MR1454264 (98h:14038)

[Kov97c] S. J. KOVÁCS: *Relative de Rham complex for non-smooth morphisms*, Birational algebraic geometry (Baltimore, MD, 1996), Contemp. Math., vol. 207, Amer. Math. Soc., Providence, RI, 1997, pp. 89–100. MR1462927 (98f:14014)

[Kov99] S. J. KOVÁCS: *Rational, log canonical, Du Bois singularities: on the conjectures of Kollár and Steenbrink*, Compositio Math. **118** (1999), no. 2, 123–133. MR1713307 (2001g:14022)

[Kov00a] S. J. KOVÁCS: *Algebraic hyperbolicity of fine moduli spaces*, J. Algebraic Geom. **9** (2000), no. 1, 165–174. MR1713524 (2000i:14017)

[Kov00b] S. J. KOVÁCS: *A characterization of rational singularities*, Duke Math. J. **102** (2000), no. 2, 187–191. MR1749436 (2002b:14005)

[Kov00c] S. J. KOVÁCS: *Rational, log canonical, Du Bois singularities. II. Kodaira vanishing and small deformations*, Compositio Math. **121** (2000), no. 3, 297–304. MR1761628 (2001m:14028)

[Kov02] S. J. KOVÁCS: *Logarithmic vanishing theorems and Arakelov-Parshin boundedness for singular varieties*, Compositio Math. **131** (2002), no. 3, 291–317. MR1905025 (2003a:14025)

[Kov03a] S. J. KOVÁCS: *Families of varieties of general type: the Shafarevich conjecture and related problems*, Higher dimensional varieties and rational points (Budapest, 2001), Bolyai Soc. Math. Stud., vol. 12, Springer, Berlin, 2003, pp. 133–167. MR2011746 (2004j:14041)

[Kov03b] S. J. KOVÁCS: *Vanishing theorems, boundedness and hyperbolicity over higher-dimensional bases*, Proc. Amer. Math. Soc. **131** (2003), no. 11, 3353–3364 (electronic). MR1990623 (2004f:14047)

[Kov03c] S. J. KOVÁCS: *Viehweg's conjecture for families over $\mathbb{P}^n$*, Comm. Algebra **31** (2003), no. 8, 3983–3991, Special issue in honor of Steven L. Kleiman. MR2007392 (2004h:14038)

[Kov05a] S. J. KOVÁCS: *Spectral sequences associated to morphisms of locally free sheaves*, Recent progress in arithmetic and algebraic geometry, Contemp. Math., vol. 386, Amer. Math. Soc., Providence, RI, 2005, pp. 57–85. MR2182771 (2006j:14016)

[Kov05b] S. J. KOVÁCS: *Strong non-isotriviality and rigidity*, Recent progress in arithmetic and algebraic geometry, Contemp. Math., vol. 386, Amer. Math. Soc., Providence, RI, 2005, pp. 47–55. MR2182770 (2006i:14008)

[Kov09] S. J. KOVÁCS: *Subvarieties of moduli stacks of canonically polarized varieties: generalizations of Shafarevich's conjecture*, Algebraic geometry—Seattle 2005. Part 2, Proc. Sympos. Pure Math., vol. 80, Amer. Math. Soc., Providence, RI, 2009, pp. 685–709. MR2483952

[KL06] S. J. KOVÁCS AND M. LIEBLICH: *Boundedness of families of canonically polarized manifolds: A higher dimensional analogue of Shafarevich's conjecture*, Ann. Math., to appear. arXiv:math.AG/0611672

[KS09] S. J. KOVÁCS AND K. SCHWEDE: *Hodge theory meets the minimal model program: a survey of log canonical and Du Bois singularities*, 2009. arXiv:0909.0993v1 [math.AG]

[KSS10] S. J. KOVÁCS, K. E. SCHWEDE, AND K. E. SMITH: *The canonical sheaf of Du Bois singularities*, to appear in Advances in Math. (2010). DOI: 10.1016/j.aim.2010.01.020

[Lan87] S. LANG: *Introduction to complex hyperbolic spaces*, Springer-Verlag, New York, 1987. MR886677 (88f:32065)

[Lan97] S. LANG: *Survey of diophantine geometry*, Springer-Verlag, New York, 1997.

[Laz04a] R. LAZARSFELD: *Positivity in algebraic geometry. I*, Ergebnisse der Mathematik und ihrer Grenzgebiete. 3. Folge. A Series of Modern Surveys in Mathematics [Results in Mathematics and Related Areas. 3rd Series], vol. 48, Springer-Verlag, Berlin, 2004, Classical setting: line bundles and linear series. MR2095471 (2005k:14001a)

[Laz04b] R. LAZARSFELD: *Positivity in algebraic geometry. II*, Ergebnisse der Mathematik und ihrer Grenzgebiete. 3. Folge. A Series of Modern Surveys in Mathematics [Results in Mathematics and Related Areas. 3rd Series], vol. 49, Springer-Verlag, Berlin, 2004, Positivity for vector bundles, and multiplier ideals. MR2095472 (2005k:14001b)

[Laz09] R. LAZARSFELD: *A short course on multiplier ideals*, 2009. arXiv:0901.0651v1 [math.AG]

[Laz08] V. LAZIĆ: *Towards finite generation of the canonical ring without the mmp*, 2008.

[LTYZ05] K. LIU, A. TODOROV, S.-T. YAU, AND K. ZUO: *Shafarevich's conjecture for CY manifolds. I*, Q. J. Pure Appl. Math. **1** (2005), no. 1, 28–67. MR2155142

[Loo95] E. LOOIJENGA: *On the tautological ring of $\mathscr{M}_g$*, Invent. Math. **121** (1995), no. 2, 411–419. MR1346214 (96g:14021)

[Man63] J. I. MANIN: *Rational points on algebraic curves over function fields*, Izv. Akad. Nauk SSSR Ser. Mat. **27** (1963), 1395–1440. MR0157971 (28 #1199)

[Mat02] K. MATSUKI: *Introduction to the Mori program*, Universitext, Springer-Verlag, New York, 2002. MR1875410 (2002m:14011)

[Mat72] T. MATSUSAKA: *Polarized varieties with a given Hilbert polynomial*, Amer. J. Math. **94** (1972), 1027–1077. MR0337960 (49 #2729)

[MM64] T. MATSUSAKA AND D. MUMFORD: *Two fundamental theorems on deformations of polarized varieties*, Amer. J. Math. **86** (1964), 668–684. MR0171778 (30 #2005)

[McK07] J. MCKERNAN: *Confined divisors*, Flips for 3-folds and 4-folds, Oxford Lecture Ser. Math. Appl., vol. 35, Oxford Univ. Press, Oxford, 2007, pp. 121–133. MR2359345

[MV07] J. D. McNeal and D. Varolin: *Analytic inversion of adjunction: L^2 extension theorems with gain*, Ann. Inst. Fourier (Grenoble) **57** (2007), no. 3, 703–718. MR2336826 (2008h:32012)

[Mig95] L. Migliorini: *A smooth family of minimal surfaces of general type over a curve of genus at most one is trivial*, J. Algebraic Geom. **4** (1995), no. 2, 353–361. MR1311355 (95m:14023)

[Mil86] E. Y. Miller: *The homology of the mapping class group*, J. Differential Geom. **24** (1986), no. 1, 1–14. MR857372 (88b:32051)

[Miy88] Y. Miyaoka: *Abundance conjecture for 3-folds: case $\nu = 1$*, Compositio Math. **68** (1988), no. 2, 203–220. MR966580 (89m:14023)

[MVZ05] M. Moeller, E. Viehweg, and K. Zuo: *Special families of curves, of Abelian varieties, and of certain minimal manifolds over curves*, 2005. arXiv:math.AG/0512154

[Mor82] S. Mori: *Threefolds whose canonical bundles are not numerically effective*, Ann. of Math. (2) **116** (1982), no. 1, 133–176. MR662120 (84e:14032)

[Mor87] S. Mori: *Classification of higher-dimensional varieties*, Algebraic geometry, Bowdoin, 1985 (Brunswick, Maine, 1985), Proc. Sympos. Pure Math., vol. 46, Amer. Math. Soc., Providence, RI, 1987, pp. 269–331. MR927961 (89a:14040)

[Mor88] S. Mori: *Flip theorem and the existence of minimal models for 3-folds*, J. Amer. Math. Soc. **1** (1988), no. 1, 117–253. MR924704 (89a:14048)

[Mor86] D. R. Morrison: *A remark on: "On the plurigenera of minimal algebraic 3-folds with $K \equiv 0$" [Math. Ann. **275** (1986), no. 4, 539–546; MR0859328 (88c:14049)] by Y. Kawamata*, Math. Ann. **275** (1986), no. 4, 547–553. MR859329 (88c:14050)

[Mum66] D. Mumford: *Lectures on curves on an algebraic surface*, With a section by G. M. Bergman. Annals of Mathematics Studies, No. 59, Princeton University Press, Princeton, N.J., 1966. MR0209285 (35 #187)

[Mum70] D. Mumford: *Abelian varieties*, Tata Institute of Fundamental Research Studies in Mathematics, No. 5, Published for the Tata Institute of Fundamental Research, Bombay, 1970. MR0282985 (44 #219)

[Nag62] M. Nagata: *Imbedding of an abstract variety in a complete variety*, J. Math. Kyoto Univ. **2** (1962), 1–10. MR0142549 (26 #118)

[Nak04] N. Nakayama: *Zariski-decomposition and abundance*, MSJ Memoirs, vol. 14, Mathematical Society of Japan, Tokyo, 2004. MR2104208 (2005h:14015)

[Nav88] V. Navarro Aznar: *Théorèmes d'annulation*, Hyperrésolutions cubiques et descente cohomologique (Barcelona, 1982), Lecture Notes in Math., vol. 1335, Springer, Berlin, 1988, pp. 133–160. MR972988

[OV01] K. OGUISO AND E. VIEHWEG: *On the isotriviality of families of elliptic surfaces*, J. Algebraic Geom. **10** (2001), no. 3, 569–598. MR1832333 (2002d:14054)

[Oor74] F. OORT: *Subvarieties of moduli spaces*, Invent. Math. **24** (1974), 95–119. MR0424813 (54 #12771)

[Oor95] F. OORT: *Complete subvarieties of moduli spaces*, Abelian varieties (Egloffstein, 1993), de Gruyter, Berlin, 1995, pp. 225–235. MR1336609 (96e:14026)

[Par68] A. N. PARSHIN: *Algebraic curves over function fields. I*, Izv. Akad. Nauk SSSR Ser. Mat. **32** (1968), 1191–1219. MR0257086 (41 #1740)

[Pău07] M. PĂUN: *Siu's invariance of plurigenera: a one-tower proof*, J. Differential Geom. **76** (2007), no. 3, 485–493. MR2331528 (2008h:32025)

[Pău08] M. PĂUN: *Relative critical exponents, non-vanishing and metrics with minimal singularities*, 2008. arXiv:0807.3109v1 [math.CV]

[PS08] C. A. M. PETERS AND J. H. M. STEENBRINK: *Mixed Hodge structures*, Ergebnisse der Mathematik und ihrer Grenzgebiete. 3. Folge. A Series of Modern Surveys in Mathematics [Results in Mathematics and Related Areas. 3rd Series], vol. 52, Springer-Verlag, Berlin, 2008. MR2393625

[Ram72] C. P. RAMANUJAM: *Remarks on the Kodaira vanishing theorem*, J. Indian Math. Soc. (N.S.) **36** (1972), 41–51. MR0330164 (48 #8502)

[Rei87] M. REID: *Young person's guide to canonical singularities*, Algebraic geometry, Bowdoin, 1985 (Brunswick, Maine, 1985), Proc. Sympos. Pure Math., vol. 46, Amer. Math. Soc., Providence, RI, 1987, pp. 345–414. MR927963 (89b:14016)

[Rei97] M. REID: *Chapters on algebraic surfaces*, Complex algebraic geometry (Park City, UT, 1993), IAS/Park City Math. Ser., vol. 3, Amer. Math. Soc., Providence, RI, 1997, pp. 3–159. MR1442522 (98d:14049)

[Sat56] I. SATAKE: *On the compactification of the Siegel space*, J. Indian Math. Soc. (N.S.) **20** (1956), 259–281. MR0084842 (18,934c)

[Sch07] K. SCHWEDE: *A simple characterization of Du Bois singularities*, Compos. Math. **143** (2007), no. 4, 813–828. MR2339829

[Sch08] K. SCHWEDE: *Centers of F-purity*, 2008. arXiv:0807.1654v3 [math.AC]

[ST07] K. SCHWEDE AND S. TAKAGI: *Rational singularities associated to pairs*, 2007. arXiv:0708.1990v4 [math.AG]

[Sha63] I. R. SHAFAREVICH: *Algebraic number fields*, Proc. Internat. Congr. Mathematicians (Stockholm, 1962), Inst. Mittag-Leffler, Djursholm, 1963, pp. 163–176, English translation: Amer. Math. Soc. Transl. (2) **31** (1963), 25–39. MR0202709 (34 #2569)

[Sho96] V. V. SHOKUROV: *3-fold log models*, J. Math. Sci. **81** (1996), no. 3, 2667–2699, Algebraic geometry, 4. MR1420223 (97i:14015)

[Sho03] V. V. SHOKUROV: *Prelimiting flips*, Tr. Mat. Inst. Steklova **240** (2003), no. Biratsion. Geom. Linein. Sist. Konechno Porozhdennye Algebry, 82–219. MR1993750 (2004k:14024)

[Sho04] V. V. SHOKUROV: *Letters of a bi-rationalist. V. Minimal log discrepancies and termination of log flips*, Tr. Mat. Inst. Steklova **246** (2004), 328–351. MR2101303 (2006b:14019)

[Sho06] V. V. SHOKUROV: *Letters of a bi-rationalist: VII. ordered termination*, 2006. arXiv:0607.822v2 [math.AG]

[Siu98] Y.-T. SIU: *Invariance of plurigenera*, Invent. Math. **134** (1998), no. 3, 661–673. MR1660941 (99i:32035)

[Siu04] Y.-T. SIU: *Invariance of plurigenera and torsion-freeness of direct image sheaves of pluricanonical bundles*, Finite or infinite dimensional complex analysis and applications, Adv. Complex Anal. Appl., vol. 2, Kluwer Acad. Publ., Dordrecht, 2004, pp. 45–83. MR2058399 (2005j:32020)

[Smi97] K. E. SMITH: *Vanishing, singularities and effective bounds via prime characteristic local algebra*, Algebraic geometry—Santa Cruz 1995, Proc. Sympos. Pure Math., vol. 62, Amer. Math. Soc., Providence, RI, 1997, pp. 289–325. MR1492526 (99a:14026)

[Ste85] J. H. M. STEENBRINK: *Vanishing theorems on singular spaces*, Astérisque (1985), no. 130, 330–341, Differential systems and singularities (Luminy, 1983). MR804061 (87j:14026)

[Ste83] J. H. M. STEENBRINK: *Mixed Hodge structures associated with isolated singularities*, Singularities, Part 2 (Arcata, Calif., 1981), Proc. Sympos. Pure Math., vol. 40, Amer. Math. Soc., Providence, RI, 1983, pp. 513–536. MR713277 (85d:32044)

[Sza94] E. SZABÓ: *Divisorial log terminal singularities*, J. Math. Sci. Univ. Tokyo **1** (1994), no. 3, 631–639. MR1322695 (96f:14019)

[Tak06] S. TAKAYAMA: *Pluricanonical systems on algebraic varieties of general type*, Invent. Math. **165** (2006), no. 3, 551–587. MR2242627 (2007m:14014)

[Tak07] S. TAKAYAMA: *On the invariance and the lower semi-continuity of plurigenera of algebraic varieties*, J. Algebraic Geom. **16** (2007), no. 1, 1–18. MR2257317 (2007i:32021)

[Tsu04] H. TSUJI: *Subadjunction theorem*, Complex analysis in several variables – Memorial Conference of Kiyoshi Oka's Centennial Birthday, Adv. Stud. Pure Math., vol. 42, Math. Soc. Japan, Tokyo, 2004, pp. 313–318. MR2087064 (2005j:32010)

[Tsu07] H. TSUJI: *Pluricanonical systems of projective varieties of general type. II*, Osaka J. Math. **44** (2007), no. 3, 723–764. MR2360948

[Var08] D. VAROLIN: *A Takayama-type extension theorem*, Compos. Math. **144** (2008), no. 2, 522–540. MR2406122

[Vie82] E. VIEHWEG: *Vanishing theorems*, J. Reine Angew. Math. **335** (1982), 1–8. MR667459 (83m:14011)

[Vie83a] E. VIEHWEG: *Weak positivity and the additivity of the Kodaira dimension for certain fibre spaces*, Algebraic varieties and analytic varieties (Tokyo, 1981), Adv. Stud. Pure Math., vol. 1, North-Holland, Amsterdam, 1983, pp. 329–353. MR715656 (85b:14041)

[Vie83b] E. VIEHWEG: *Weak positivity and the additivity of the Kodaira dimension. II. The local Torelli map*, Classification of algebraic and analytic manifolds (Katata, 1982), Progr. Math., vol. 39, Birkhäuser Boston, Boston, MA, 1983, pp. 567–589. MR728619 (85i:14020)

[Vie89] E. VIEHWEG: *Weak positivity and the stability of certain Hilbert points*, Invent. Math. **96** (1989), no. 3, 639–667. MR996558 (90i:14037)

[Vie90a] E. VIEHWEG: *Weak positivity and the stability of certain Hilbert points. II*, Invent. Math. **101** (1990), no. 1, 191–223. MR1055715 (91f:14032)

[Vie90b] E. VIEHWEG: *Weak positivity and the stability of certain Hilbert points. III*, Invent. Math. **101** (1990), no. 3, 521–543. MR1062794 (92f:32033)

[Vie91] E. VIEHWEG: *Quasi-projective quotients by compact equivalence relations*, Math. Ann. **289** (1991), no. 2, 297–314. MR1092177 (92d:14028)

[Vie95] E. VIEHWEG: *Quasi-projective moduli for polarized manifolds*, Ergebnisse der Mathematik und ihrer Grenzgebiete (3) [Results in Mathematics and Related Areas (3)], vol. 30, Springer-Verlag, Berlin, 1995. MR1368632 (97j:14001)

[Vie01] E. VIEHWEG: *Positivity of direct image sheaves and applications to families of higher dimensional manifolds*, School on Vanishing Theorems and Effective Results in Algebraic Geometry (Trieste, 2000), ICTP Lect. Notes, vol. 6, Abdus Salam Int. Cent. Theoret. Phys., Trieste, 2001, pp. 249–284. MR1919460 (2003f:14024)

[Vie06] E. VIEHWEG: *Compactifications of smooth families and of moduli spaces of polarized manifolds*, 2006. arXiv:math.AG/0605093

[VZ01] E. VIEHWEG AND K. ZUO: *On the isotriviality of families of projective manifolds over curves*, J. Algebraic Geom. **10** (2001), no. 4, 781–799. MR1838979 (2002g:14012)

[VZ02] E. VIEHWEG AND K. ZUO: *Base spaces of non-isotrivial families of smooth minimal models*, Complex geometry (Göttingen, 2000), Springer, Berlin, 2002, pp. 279–328. MR1922109 (2003h:14019)

[VZ03a] E. VIEHWEG AND K. ZUO: *Discreteness of minimal models of Kodaira dimension zero and subvarieties of moduli stacks*, Surveys in differential geometry, Vol. VIII (Boston, MA, 2002), Surv. Differ. Geom., VIII, Int. Press, Somerville, MA, 2003, pp. 337–356. MR2039995 (2005d:14018)

[VZ03b] E. VIEHWEG AND K. ZUO: *On the Brody hyperbolicity of moduli spaces for canonically polarized manifolds*, Duke Math. J. **118** (2003), no. 1, 103–150. MR1978884 (2004h:14042)

[VZ05a] E. VIEHWEG AND K. ZUO: *Complex multiplication, Griffiths-Yukawa couplings, and rigidity for families of hypersurfaces*, J. Algebraic Geom. **14** (2005), no. 3, 481–528. MR2129008 (2005m:14059)

[VZ05b] E. VIEHWEG AND K. ZUO: *Geometry and arithmetic of non-rigid families of Calabi-Yau 3-folds; questions and examples*, 2005. arXiv:math.AG/0503337

[VZ05c] E. VIEHWEG AND K. ZUO: *Special subvarieties of $\mathscr{A}_g$*, 2005. arXiv:math.AG/0509037

[VZ06] E. VIEHWEG AND K. ZUO: *Numerical bounds for semi-stable families of curves or of certain higher-dimensional manifolds*, J. Algebraic Geom. **15** (2006), no. 4, 771-791. MR2237270 (2007d:14019)

[Wil95a] A. WILES: *Modular elliptic curves and Fermat's last theorem*, Ann. of Math. (2) **141** (1995), no. 3, 443–551. MR1333035 (96d:11071)

[Wil95b] A. WILES: *Modular forms, elliptic curves, and Fermat's last theorem*, Proceedings of the International Congress of Mathematicians, Vol. 1, 2 (Zürich, 1994) (Basel), Birkhäuser, 1995, pp. 243–245. MR1403925 (97f:11041)

[YT06] S. YUM-TONG: *A general non-vanishing theorem and an analytic proof of the finite generation of the canonical ring.*, 2006. arXiv:0610740 [math.AG]

[Zaa95] C. G. ZAAL: *Explicit complete curves in the moduli space of curves of genus three*, Geom. Dedicata **56** (1995), no. 2, 185–196. MR1338958 (96k:14020)

[Zaa99] C. G. ZAAL: *A complete surface in M_6 in characteristic* > 2, Compositio Math. **119** (1999), no. 2, 209–212. MR1723129 (2000i:14031)

[Zuo00] K. ZUO: *On the negativity of kernels of Kodaira-Spencer maps on Hodge bundles and applications*, Asian J. Math. **4** (2000), no. 1, 279–301, Kodaira's issue. MR1803724 (2002a:32011)

Index

Symbols

W

Z

Oberwolfach Seminars (OWS)

BIRKHÄUSER

The workshops organized by the *Mathematisches Forschungsinstitut Oberwolfach* are intended to introduce students and young mathematicians to current fields of research. By means of these well-organized seminars, also scientists from other fields will be introduced to new mathematical ideas. The publication of these workshops in the series *Oberwolfach Seminars* (formerly *DMV seminar*) makes the material available to an even larger audience.

OWS 41: Hacon, C.D. / Kovács, S., Classification of Higher Dimensional Algebraic Varieties (2010). ISBN 978-3-0346-0289-1

OWS 40: Baum, H. / Juhl, A., Conformal Differential Geometry. Q-Curvature and Conformal Holonomy (2010).
ISBN 978-3-7643-9908-5

Conformal invariants (conformally invariant tensors, conformally covariant differential operators, conformal holonomy groups etc.) are of central significance in differential geometry and physics. Well-known examples of conformally covariant operators are the Yamabe, the Paneitz, the Dirac and the twistor operator. These operators are intimately connected with the notion of Branson's Q-curvature. The aim of these lectures is to present the basic ideas and some of the recent developments around Q-curvature and conformal holonomy. The part on Q-curvature starts with a discussion of its origins and its relevance in geometry and spectral theory. The subsequent lectures describe the fundamental relation between Q-curvature and scattering theory on asymptotically hyperbolic manifolds. Building on this, they introduce the recent concept of Q-curvature polynomials and use these to reveal the recursive structure of Q-curvatures. The part on conformal holonomy starts with an introduction to Cartan connections and their holonomy groups. Then we define holonomy groups of conformal manifolds, discuss their relation to Einstein metrics and recent classification results in Riemannian and Lorentzian signature. In particular, we explain the connection between conformal holonomy and conformal Killing forms and spinors, and describe Fefferman metrics in CR geometry as Lorentzian manifolds with conformal holonomy $SU(1, m)$.

OWS 39: Drton, M. / Sturmfels, B. / Sullivant, S., Lectures on Algebraic Statistics (2008). ISBN 978-3-7643-8904-8

How does an algebraic geometer studying secant varieties further the understanding of hypothesis tests in statistics? Why would a statistician working on factor analysis raise open problems about determinantal varieties? Connections of this type are at the heart of the new field of "algebraic statistics". In this field, mathematicians and statisticians come together to solve statistical inference problems using concepts from algebraic geometry as well as related computational and combinatorial techniques. The goal of these lectures is to introduce newcomers from the different camps to algebraic statistics. The introduction will be centered around the following three observations: many important statistical models correspond to algebraic or semi-algebraic sets of parameters; the geometry of these parameter spaces determines the behaviour of widely used statistical inference procedures; computational algebraic geometry can be used to study parameter spaces and other features of statistical models.

OWS 38: Bobenko, A.I. / Schröder, P. / Sullivan, J.M. / Ziegler, G.M. (Eds.), Discrete Differential Geometry (2008). ISBN 978-3-7643-8620-7

OWS 37: Galdi, G.P. / Rannacher, R. / Robertson, A.M. / Turek, S., Hemodynamical Flows (2008). ISBN 978-3-7643-7805-9

OWS 36: Cuntz, J. / Meyer, R. / Rosenberg, J.M., Topological and Bivariant K-theory (2007). ISBN 978-3-7643-8398-5

OWS 35: Itenberg, I. / Mikhalkin, G. / Shustin, E., Tropical Algebraic Geometry (2007). ISBN 978-3-7643-8309-1

OWS 34: Lieb, E.H. / Seiringer, R. / Solovej, J.P. / Yngvason, J., The Mathematics of the Bose Gas and its Condensation (2005).
ISBN 978-3-7643-7336-8

OWS 33: Kreck, M. / Lück, W., The Novikov Conjecture: Geometry and Algebra (2004).
ISBN 978-3-7643-7141-8